AF891706

Johannes Schäuble, Patrick Jochem, Wolf Fichtner (Eds.)

Cross-border Mobility for Electric Vehicles

Selected results from one of the first cross-border field tests in Europe

Cross-border Mobility for Electric Vehicles

Selected results from one of the first cross-border field tests in Europe

edited by
Johannes Schäuble, Patrick Jochem, Wolf Fichtner

KIT Scientific Publishing

Impressum

Karlsruher Institut für Technologie (KIT)
KIT Scientific Publishing
Straße am Forum 2
D-76131 Karlsruhe

KIT Scientific Publishing is a registered trademark of Karlsruhe Institute of Technology. Reprint using the book cover is not allowed.

www.ksp.kit.edu

This document – excluding the cover, pictures and graphs – is licensed under the Creative Commons Attribution-Share Alike 3.0 DE License (CC BY-SA 3.0 DE): http://creativecommons.org/licenses/by-sa/3.0/de/

The cover page is licensed under the Creative Commons Attribution-No Derivatives 3.0 DE License (CC BY-ND 3.0 DE): http://creativecommons.org/licenses/by-nd/3.0/de/

Print on Demand 2016

ISBN 978-3-7315-0492-4
DOI 10.5445/KSP/1000052445

Acknowledgments

The content of this book is mainly based on the jointly implemented cross border field test with electric vehicles between 2011 and 2014. This was enabled by strong efforts and collaboration of our full and associated project partners and their kind and excellent representatives (cf. Section 1.1). The implementation and its accompanying research were made possible by the CROME project [ref. no. 01ME12002], funded by the Federal Ministry for Economic Affairs and Energy and the Federal Ministry of Transport and digital Infrastructure in Germany and the French program "Investissements d'avenir programme véhicule de futur". The editors would like to take this opportunity to thank Sebastian Rilling for his help to improve the content and the layout of all contributions in this book.

Gefördert durch:

Bundesministerium für Wirtschaft und Energie

Bundesministerium für Verkehr und digitale Infrastruktur

aufgrund eines Beschlusses des Deutschen Bundestags

Liberté • Égalité • Fraternité
République Française

Ministère de l'Écologie, du Développement Durable et de l'Énergie

Ministère du Redressement Productif

Ministère de l'Enseignement Supérieur et de la Recherche

Investissements d'avenir

ADEME
Direction Régionale
Franche-Comté

Table of Contents

Index of Figures

Index of Tables

List of Abbreviations

AC	Alternating current
AHC	Ascending hierarchical classification
AIFB	Institut für Angewandte Informatik und Formale Beschreibungsmethoden (Insitute of Applied Informatics and Formal Description Methods)
ALT	Alternatively needed attributes
BEV	Battery EV
CROME	Cross-border mobility for EV
CSV	Comma- or semicolon- separated values
CUMILE	Car usage model integrating long distance events
DC	Direct current
DRE	Distributed renewable energy
EC	European Commission
EDF	Électricité de France SA
EIFER	European Institute for Energy Research
EnBW	Energie Baden-Württemberg AG
ENTD	Enquête Nationale Transports et Déplacements (national survey on transports and locomotion)
EU	European Union
EV	Electric vehicle(s)
EVSE	Electric vehicle supply equipment
GPS	Global Positioning System
HTTP(S)	Hypertext Transfer Protocol (Secure)
ICAME	Innovation Commerciale, Analyse des Marchés et de leur Environnement (commercial innovation, analyse of markets and their environment)

ICEV	Internal combustion engine vehicle
IMEI	International Mobile Equipment Identity
INT	Internal attributes
ITEI	International Mobile Station Equipment Identity
JSON	JavaScript Object Notation
KHL	Keyhole Markup Language
KIT	Karlsruhe Institute of Technology
LTE	Long Term Evolution
MID	European Measurement Instruments Directive
MOP	German Mobility Panel
MOT	Other means of transport
OEM	Original equipment manufacturer
OGC	Open Geospatial Consortium
PCC	Point of common coupling
PHEV	Plug-in hybrid electric vehicle
PI	Proportional integral
PLL	Phase-locked loop
R&D	Research and development
REQ	Required attributes
RFID	Radio-Frequency Identification
RPM	Rates per minute
SEV	Self-service EV
Smart ed	Smart Fortwo electric drive
SOAP	Simple Object Access Protocol
SOC	State of charge
SQL	Structured Query Language
SSL	Secure Sockets Layer
SUP	Supporting data

THDI	Total harmonic distortion
UML	Unified Modelling Language
UMTS	Universal Mobile Telecommunications System
UTC	Universal Time Coordinated
VDA	Verband der Automobilindustrie (association of the automotive industry)
VIN	Vehicle identification number
W3C	World Wide Web Consortium
WSDL	Web Services Description Language
WTP	Willingness to pay
XML	Extensible Markup Language

Prologue

Against the background of climate change and the limited resources of fossil fuels the governments of France and Germany both have supported an accelerated market penetration of electric vehicles (EV) for several years. When using electricity from low-carbon power generation, the substitution of internal combustion engine vehicles (ICEV) by EV will not only reduce CO_2 emissions in the transport sector, but also the dependency on fossil fuels. Nowadays, the use of EV goes still along with certain disadvantages compared to ICEV which inhibit a larger market penetration of EV. Next to high costs of EV, range limitations and a relatively long battery charging duration imply flexibility losses for EV users (cf. chapter 8). In order to reduce these drawbacks, technological advances on the vehicle side as well as a user-friendly charging infrastructure and mobility services that fulfil the users' needs are required (BMVI 2014 p.45). As vast efforts were and still are made in both countries to fulfil these requirements the need for cooperation on these concerns were eminent. Therefore, the vision of the CROME (CROss-border Mobility for Electric vehicles) research project was to create and test a safe, seamless, user-friendly and reliable mobility with EV between France and Germany as a prefiguration of pan-European electromobility. One of its major aims was to formulate answers and suggestions to contribute to the European standardisation process of charging infrastructure for electromobility (such as electric supply, cables and plugs) and of corresponding services (e.g. identification, billing, and roaming) and to provide stakeholders with an early customer feedback. Several significant milestones have been achieved during the project:

- A wide-scale cross-border field demonstration of mobility with more than 100 EV was performed in 2011, introducing fully public interoperable charging stations (EVSE), ensuring easy access and charging of EV all over the French and German CROME area.

- The customer acceptance of electromobility as well as the users' needs regarding charging in the context of cross-border mobility were investigated.
- Charging services enabling simplified identification and cross-border billing and charging spot availability as well as reservation was developed, implemented and tested. This serves as a cornerstone of the European roaming concept.
- The project contributed significantly to the European standardisation process of charging infrastructure and corresponding services.
- Besides these achievements, this book provides our main findings of the accompanying research.

In the first chapter the **editors** give a brief overview on the projects context and its main characteristics presenting the involved partners, their performed tasks, the concept of interoperability, the fleet test schedule, the involved EV and the installed infrastructure.

In the second chapter **Daniel Ried** presents the data repository of the fleet test, i.e. data collection, transfer and storage. The design and architecture of the data repository is focused on a secure and reliable operation of all transfers and storage. The heterogeneity of datasets was high: Technical data with a very high timely resolution came from the vehicle (via internet from vehicle manufacturers), geo-positions and other trip related data (recorded by smartphones) were sent by data-streams, and other data from the vehicle users were merged with a unique ID system. Additionally, various interfaces, tools and services for the data repository were developed in order to improve the usability of data imports and processing of raw data as well as for querying, accessing and exporting aggregated business objects for further analytical processes.

In the third chapter **Peter Krasselt** gives an insight into public EV fast charging infrastructure and their reactive voltage support for the grid. Therefore he depicts standards and technical rules for their network connection as well as how network disturbance can be assessed, and he proposes a fast charging infrastructure topology. Finally a detailed analysis of reactive power voltage control schemes is performed and its results are highlighted.

In the fourth chapter **Magali Pierre** and **Anne-Sophie Fulda** present the results from 45 interviews performed during the fleet test, all respondents are EV users – some are employees using the EV of their company and others are private EV owners: First they take a deeper look into EV usage in a professional context, before they refer to interviews with EV users which influenced the EV purchase decision. Finally, changes in mobility patterns and experiences of EV usage during the first year are highlighted.

In the fifth chapter **Felix Vogel** and **Hartmut Schmeck** give an overview on the development of electromobility services within the fleet test and in general. Furthermore, they describe an analysis of the users' mobility behaviour with EV as well as the usability of e-services in France and in Germany and expose its results.

In chapter six **Eva Weis** and **Silvia Balaban** take a look at the harmonisation process of European legislation in the context of electromobility. They explicate the framework conditions and analyse the general ways of attaining a European harmonisation within its limits. The chapter shows the harmonisation approach of the European Commission on the example of fast charging infrastructure. Further aspects connected to the use of EV regarding their need of being harmonized are equally outlined.

In chapter seven **Mathias Pfriem, Frank Gauterin** and **Thomas Meyer** give an overview of the analysed usage of the EV on the basis of recorded trips during the field test. The authors present the data sample and their developed and applied data collection methods. Subsequently key figures of the observed mobility are presented. The charging behaviour is analysed additionally and reveals a mostly daily recharging from high states of charge to a full battery. Finally, national idiosyncrasies are presented by comparing characteristic mobility aspects of France and Germany.

In chapter eight **Axel Ensslen, Patrick Jochem, Martin Rometsch** and **Wolf Fichtner** analyse purchase intentions of EV users and fleet managers who experienced EV during a period of about two years in a professional context. Therefore, after presenting a brief literature review, they characterize the sample of an online questionnaire. A binary logistic regression model is derived from the survey data providing information on private EV purchase intentions.

In chapter nine **Elise Nimal** and **Anne-Sophie Fulda** analyse the collected data during the operation of charging stations installed in France in consideration of frequency, location and user characteristics. Additionally they have a deeper look into the user behaviour related to the use of charging infrastructure.

Tim Hilgert, Martin Kagerbauer, Christine Weiss and **Peter Vortisch** pursue the question of how EV can be used in people's everyday life in chapter ten. After a brief presentation of the underlying data, models and methods they analyse the usage of EV in the field test and give a short comparison of vehicle usage in France and Germany. Using the national household travel survey data (German Mobility Panel (MOP)) they estimate the share of German car trips that can be substituted by EV and compare this with the situation in company fleets.

Condensed key findings of the accompanying studies in the fields of charging infrastructure, electric vehicle usage, user acceptance and demand for mobility within the fleet test are described briefly in the following:

Charging infrastructure: Fast charging systems have a high influence on the voltage quality in the low-voltage network. The effect on the change in voltage is particularly high. Boundary value violations are observed in typical low-voltage networks for a distance of over 250 meters between the fast charging unit and the local substation. Active network supporting fast charge units may be operated in long branches exceeding 250 m. Here, the voltage band is stabilized by reactive power support and the maximum distance can be increased by 30%. When comparing different regulatory approaches for voltage support a hysteresis-based regulatory approach has achieved the best results in rural, suburban and urban grid area. In compliance with the normative requirements for the voltage quality the additional power losses were in this case in the range of 1-5% during the loading operation due to the additional reactive power load flow. Additional fast charging systems may attenuate fast voltage changes. Here, the control parameters of all connected consumers who provide active voltage stabilizing are to be chosen with care to avoid control oscillations and instabilities.

Vehicle use: In the field test, EV are applied mainly in commercial fleets of local focused institutions and are mainly used in urban areas within a radius of 40 km maximum. The rides are mostly short (<10 km) and also in combination of all day trips well below the maximum possible range. Nevertheless, the vehicles are charged almost daily, so the battery is used almost exclusively in the upper half of its capacity. The share of journeys over 15 km in length is significantly higher in France, because more overland trips are made, resulting in relatively faster velocity profiles.

User acceptance: EV in the context of CROME are perceived more environmentally friendly by the French respondents than by the German users. In particular, the characteristic of EV to cause low CO_2 emissions is evaluated significantly better by the French respondents. The German field test participants evaluate the factor that EV are perceived as innovative and environmentally friendly, both inside and outside their organization by third parties, significantly better than the French field test participants. CROME field test participants, who already drove a few trips with EV, are more likely to purchase an EV than those who had no experience with EV. The multi-modal data repository with user acceptance, mobility and technical data provides a good basis for analysing the applicability of EV in commercial fleets.

Replaceability of ICEV by EV: CROME survey participants have a significant consumer preference for EV and high annual mileage. Nevertheless, for the replacement of an ICEV by an EV behavioural changes are mostly necessary. Due to the car choices within a vehicle fleet the use of EV as a supplement in company vehicle fleets is a realistic scenario, as for every ride a reach optimal vehicle is available. However, we still see the requirement to harmonize and adjust the national legislation for EV.

Karlsruhe, October 2015

Johannes Schäuble, Patrick Jochem, Wolf Fichtner

1 Context of the fleet test

Johannes Schäuble, Patrick Jochem, Wolf Fichtner

French-German Institute for Environmental Research /
Institute for Industrial Production, Karlsruhe Institute of Technology

This chapter will give a short introduction to the context of the fleet test. The following chapters will then provide detailed information on the research conducted during the time of the fleet test. As each chapter solely provides information about the individual treated topic this chapter will rather highlight the environment in which these researches had been performed. This will help the reader to understand certain singularities which had to be faced during the fleet test.

In section 1.1 the involved partners and their tasks will be described briefly. In the following section we explicate the concept and the challenges of interoperability during the fleet test. Section four gives a rough overview on the fleet schedule and section 1.4 gives an overview of all employed EV. Finally the charging infrastructure of the fleet is presented.

1.1 Partners in the fleet test and their main activities

Different French and German industry partners and research organisations participated and introduced their knowledge into the fleet test. They were supported by several associated partners, mainly French and German regional institutions. The project was funded by French and German Ministries: On French side the Ministry for Ecology, Sustainable Development and Energy (Ministère de l'Ecologie, du Développement Durable et de l'Energie), the Ministry of Industry (Ministère du Redressement Productif) and the Ministry of Academic Education and Research (Ministère de l'Enseignement Supérieur et de la Recherche), on the German side the Federal Ministry for Economic Affairs and Energy (Bundesministerium für

Wirtschaft und Energie) and the Federal Ministry of Transport and digital Infrastructure (Bundesministerium für Verkehr und digitale Infrastruktur).

In the following a short description of the main tasks of the full partners is given. We are focusing here only on the tasks related to the fleet test and do not mention all other tasks undertaken by the partners. Furthermore, we make no claim to be complete here.

The **original equipment manufacturers (OEM)** Daimler AG, Dr. Ing. h.c. F. Porsche AG, PSA Peugeot Citroën SA and Renault SA provided the EV which were employed in the project. Their specifications are described in section five. Daimler developed a conformance test tool for Mode 3 charging processes (ISO 15118), accompanied the cross-border infrastructure standardisation process and also brought two types of EV into the project. A total of 60 cars split between the models smart fortwo electric drive and Mercedes-Benz A-Klasse E-CELL were involved in the project. Hereby, the company could gain some important insight on market research and customer acceptance. Within the fleet test Porsche focused on the installation and the actual testing of battery-operated electric sports cars without combustion engine in form of the Porsche Boxster e model as well as the testing of vehicles from the Panamera family with plug-in technologies and their use in cross-border mobility. The brands Peugeot and Citroën participated in the fleet test with approximately 35 mass produced EV: About 30 Peugeot iOn and Citroën C-ZERO as well as some Peugeot Partner Origin and Citroën Berlingo were brought into the fleet test and used in normal conditions by private and commercial clients. In cooperation with the research institute IFSTTAR ten Peugeot iOn and Citroën C-ZERO were equipped with additional data loggers. Renault brought seven mass-produced Kangoo Z.E. into the fleet.

For all OEMs the main findings were primarily derived from the challenge of a cross-border use of EV, as e.g. the different mode of operation of charging stations and several billing and payment systems as well as different plug types. The different additionally aspects that were put under test included the batteries' durability, charging and discharging cycles, the users' acceptance EV as well as the safety in traffic under everyday conditions.

The **utilities** Électricité de France SA (EDF) and Energie Baden-Württemberg AG (EnBW) provided and developed several EV specific services (cf. chapter 5) as well as comprehensive interoperable electric vehicle supply equipment (EVSE) infrastructure (cf. section 1.5). EDF focused within the fleet test on (1) user acceptance analysis based on experiences during former projects (e.g. Kléber in Strasbourg), (2) designing and defining studies on the interoperable charging infrastructures and roaming platform as well as (3) supporting the local territories (Communauté urbaine de Strasbourg, region Moselle, Thionville, Forbach, Sarreguemines, and Colmar) in order to deploy public charging stations (e.g. upstream specifications). For the latter task European conform calls for tenders for interoperable EVSE were developed, which were used as a blueprint in many other regions during the last years. Additionally, together with NISSAN and CORA CHAdeMO fast charging EVSE where installed and operated. EnBW took over the corresponding role on the German side and installed and are operating an interoperable charging infrastructure in Baden along the French-German border in cooperation with local energy suppliers as well as the development of innovative tariff and access concepts for public EVSE. Whereas in Germany the rollout of the EVSE was supported by EnBW, in France the rollout was financed by the local authorities (which seems to be closer to real market conditions). For the rollout EnBW made use of the experiences from the MeRegioMobil project in Karlsruhe and Stuttgart. Together with other partners EDF and EnBW contributed substantially to the development and implementation of the interoperable electromobility platform.

The **electric infrastructure and service oriented industrial partners** such as Schneider Electric SA, Siemens AG and Bosch Software Innovations GmbH installed further infrastructure and developed a comprehensive roaming layer, which was used to provide service interoperability in the fleet test. This infrastructure represents a cornerstone of current European roaming for EV by inertercharge (cf. chapter 5). Bosch Software Innovations provided and operated special internet services for electromobility. These services allowed the necessary communication among all parties involved in the fleet test to offer an interoperable charging for the customers. Furthermore, other added value services such as navigation and networked fleet management were discussed with all partners and partly implemented. The Siemens AG supported the project partners in the topic of interoperability of the alternating current (AC) charging infrastructure (Mode 3), security aspects during the charging process and the installation and testing of direct current (DC) charging stations (Mode 4, CHAdeMO). Additionally, Siemens integrated infrastructure in the developed roaming layer and tested existing standards on plugs and charging protocols between the EV and the EVSE (mainly ISO15118) as well as between EVSE and the backend (e.g. OCPP). Schneider Electric SA performed different tasks within the fleet test (e.g. requirements for interoperable charging stations deployed in France), providing Mode 3 charging solutions, contributing to the standardization process on charging interfaces as well as designing and developing advanced energy management systems for EVSE.

The core task of the Institut français des sciences et technologies des transports, de l'aménagement et des réseaux (IFSTTAR) was to analyse vehicle data from the french EV. The Karlsruhe Institute of Technology (KIT) and the European Institute for Energy Research (EIFER) carried out **accompanying research analyses** during the whole project duration. The research performed by these institutes cover most scientific aspects within the fleet test and is described in detail in this book.

Associated partners are mainly regional partners on both sides of the border. On the French side these are regional corporations:

- Conseil Général de la Moselle,
- Communauté Urbaine de Strasbourg,
- Région Alsace,

on the German side the regional energy suppliers:

- E-Werk Mittelbaden,
- Stadtwerke Karlsruhe,
- Stadtwerke Baden-Baden and
- Star.Energiewerk Rastatt.

Further partners are:

- Verband der Automobilindustrie (VDA) and
- EIFER.

1.2 Interoperability

Interoperability was identified as a key aspect within the bi-national fleet test. In the beginning of the fleet test three main domains were identified in which certain interoperability had to be established in order to allow cross border mobility for EV: (i) hardware (in particular for EV charging), (ii) software and services as well as (iii) billing (for charging and service usage). While these domains usually require analysis in the field of systems engineering and information technology to allow the necessary information exchange we integrated organisational, political, and social perspectives, too. In doing so, a coherent infrastructure of technically differing solutions and services for the individual components in the fleet test were developed and implemented.The objective was to provide and test a first interoperable system of several dozen EVSE in two countries which may serve as a cornerstone for a barrier-free Europe-wide interoperable system, which allows charging an EV wherever needed.

Therefore, the following two foci (minimum requirements) were prioritized in the fleet test: each EV can technically connect to all EVSE within the system (i.e. compliant socket system) and each EV user can authenticate oneself at each EVSE (e.g. via Radio-Frequency Identification (RFID) card).

At the beginning of the fleet test in 2011, the **hardware interoperability** was not given. On the French side most charging stations (mainly operated during the Kléber project) were equipped with a domestic socket (Mode 2) and for Mode 3 with the Italian-French socket type (Type 3 plug system). Furthermore, several charging stations provided a "camping socket" outlet (Blue P+N+E-Socket according IEC 60309) for the first TOYOTA Prius plug-in hybrid electric vehicle (PHEV) version. On the German side the MeRegioMobil based charging stations are equipped with domestic sockets and Type 2 (i.e. "Mennekes") outlets. Hence, only an interoperable Mode 2 charging was possible and not the preferred Mode 3 charging.

Therefore, in the fleet test, the following steps had been performed: At least (additionally to the already existing infrastructure) 25 charging stations have been installed in public areas on each side of the border. Besides domestic socket outlets, these charging stations were each equipped with both, Type 2 and Type 3 socket outlets ("dual type socket" charging) and allowed for up to 22 kW charging. It was intended to retrofit these dual socket spots later in order to comply with the European standardisation decision[1]. This was taken in 2013 and favours the Type 2 socket system.

In 2011 the situation of **Software, service and billing interoperability** has been even worse. Most of the charging processes were free of charge and authentication technologies – if necessary – were very diverse. Only the already installed EVSE by EnBW provided a unique billing system.

For improving this situation a new and broadly accepted concept has to be developed. This concept should consider requirements of all European stakeholders (i.e. EVSE as well as EV providers and operators, utilities, etc.) as well as EV users. Also future developments in specific services in all relevant fields (e.g. navigation, EVSE reservation, further authentication) should be taken into account. To assure interoperability for software and services the following steps have been established during the fleet test: (1) unified access and authentication concepts based on proofed systems like

1 Directive COM(2013)0018

RFID, NFC, (2) cross border billing and roaming concepts based on different payment methods, (3) interoperable smart charging concepts (based on IEC 15118), (4) management concepts to control and monitor the charging infrastructure as well as (5) unified connectors for data exchange between the partners are analysed developed and implemented. Basis for this concept was a unified roaming service layer (cf. Figure 2) which also served as basis for several implemented smart functionalities e.g. route optimizing and charging spot reservation.

Figure 1: Dual Charging Station for Type 1 and 2 in France.

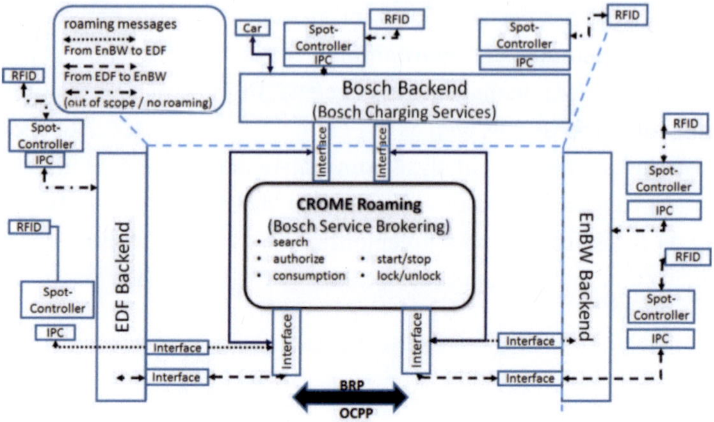

Figure 2: The roaming layer architecture by Bosch used in CROME in order to allow interoperable charging and other services.

1.3 Timeline

From January 2011 to December 2014 the following steps have been performed during the fleet test:

In the beginning of 2011 the EV prototypes were set up, commercial conditions fixed and possible users identified[2]. Afterwards the roll out of the EV started and their usage was monitored until the end of 2013.

In the beginning of 2011 first the technical specifications for the EVSE had to be validated in order to subsequently install and operate the dual type EVSE within the fleet test region. In the beginning of 2012 EVSE with DC-fast-charging capabilities for EV (Mode 4, CHAdeMO) were installed and operated[3]. Additionally, German OEM tested DC charging (Mode 4) with CCS-System with their company cars in 2013.

2 The users were neither chosen nor subsidized. This resulted in a user characteristic which is specified in chapters 8 and 10.

3 This infrastructure has also been used for the Alsace Corridor Energétique and the project RheinMobil.

Regarding services from 2011 to 2012 multiple tasks concerning the communication of EVSE and EV, the unified access and authentication system as well as the development of the roaming service layer were performed. In 2013 further advanced services and the connection of the EVSE to the backend system were implemented and operated.

A billing system between the German EVSE operators in the fleet test was established in 2012. In 2013 additionally French EVSE operators were included into the billing system.

The conducted accompanying research exemplified in detail in the following chapters started with preliminary tasks in 2011 and continued until 2014 when a final deep analysis of the data from the fleet test has been conducted.

1.4 EV in the fleet test

The target number of cars to be brought into the fleet test was set to a minimum of 100 of which most were battery electric vehicles (BEV) and only few PHEV. The fleet included vehicles ranging from small cars to sports cars and utility vehicles. These were either series vehicles which are and were offered for sale to final customers or small series vehicles which could be leased by the customers (see chapter seven and eight for a detailed discussion on the customers). Within the fleet test even some prototypes were tested by Porsche. Most EV were able to charge in Mode 2 (domestic sockets) and in Mode 3 (with Type 2 or Type 3 cables) (cf. Table 1). Only few vehicles additionally supported DC fast charging up to 43 kW.

1.5 Installed infrastructure

A key objective of the fleet test was to demonstrate solutions which allow users to connect their EV to EVSE throughout Europe.Therefore, in a first step these solutions had to take into account the infrastructure which had already been deployed in France and Germany in 2011. In the beginning of the field test each user had to carry three cords for assuring interoperability (i.e. Mode 2 with domestic plug as well as two Mode 3 cables, one with a Type 3 and the other one with a Type 2 plug). The idea of providing each user with three different cords was abandoned as it would neither support a sustainable interoperability on charging spots nor meet the customer needs. The dual type socket solution has been introduced instead (see above) and all already operated EVSE by project partners have been altered correspondingly. They allow for Mode 2 charging and Mode 3 charging up to 22 kW. At the end of the fleet test 50 EVSE equipped with dual type sockets granted a minimum of interoperability testing the project region (cf. Figure 3).

Each EV has been delivered to the fleet test with a set of two cords: one for Mode 2 charging (domestic plug) and the other for Mode 3 charging (in Germany Type 2 plug and in France Type 3 plug).

Figure 3: Geographical venue of the fleet test.

Table 1: Characteristics of applied EV in the fleet test.

Max. speed (km/h)	Range (km)	Plug type (EV side)	Charging mode	Battery energy (kWh)	Commercial launch	Production	Category	
150	255	2	2/3	36	2010	Series	BEV	**Mercedes Benz A-Klasse E-CELL**
130	150	1	2/3/4	16,3	2010	Mass	BEV	**Peugeot iOn / Citroën C-ZERO**
110	120	2	2/3/4	23,5	2010	Mass	BEV	**Peugeot Partner Origin / Citroën Berlingo First**
150/200	170	2	2/3	26	-	Prototype	BEV	**Porsche Boxster e**
130	170	2	2/3	22	2011	Mass	BEV	**Renault Kangoo Z. E.**
100	135	2	2/3	16,5	2009	Series	BEV	**smart fortwo electric drive**
180	25	1	2/3	4,4	2012	Series	PHEV	**Toyota Prius**
270	30	2	2/3	7,5	-	Prototype	PHEV	**Porsche Panamera S Hybrid**

1.6 The project's major learnings

Europe-wide standards for the further development of electromobility are crucial

The project has demonstrated that interoperability is feasible based on the existing technologies on both sides of the border: The fleet test showed that both the type 2 and type 3 plugs and sockets can be implemented in one charging station according to the regulatory constraints on each side of the border in a way that easy retrofitting is ensured. Both types enabled charging of the EV in the project. Although technically feasible, this solution does not appear meaningful on the long term. In order to reduce the costs for the deployment of the infrastructure, complexity of hardware, and to increase the user acceptance, the CROME partners recommended an agreement at European level on one standard type of plug and the deployment of charging stations which can be easily retrofitted if decided by the infrastructure owner, so that the costs for the adaptation to a future standard remain as low as possible. Thanks to the development of common specifications for mode 3 compliant public charging stations and to the preparation of the corresponding initial call for tender together with CUS, CROME contributed to simplify the erection of mode 3 charging stations in France. The CROME terms of reference for the charging infrastructure have already been adopted by further border regions in France, e.g. Pas-de-Calais.

Fast charging is used and meets EV customer needs

The CROME project demonstrated that electromobility corridors are an appropriate pattern in order to interlink already "electrified regions". However, interoperable and multi-standard *fast charging facilities* are required in order to cope with the different EV types on the market and foster the exchange between the regions. The CROME project proved that CCS charging and high level communication according to ISO / IEC 15118 work properly.

CROME demonstrated service interoperability by roaming of services

The detailed definition added to the mode 3 specifications enabled a reliable cross-border charging. In this respect, no further developments are needed. The adoption of the CROME terms of reference for charging in mode 3 has contributed to the development of an industrial offer in terms of infrastructure.

The RFID card is a suitable media for ensuring roaming; within the project, the technology has proved to be user-friendly and reliable. In addition, a live retrieval of the information needed between the backends avoids keeping data in all the systems. Consequently, e. g. in case a RFID-card gets lost, it is sufficient to disable it in one of the systems to have it immediately disabled in all the network of connected systems.

The Open Charge Point Protocol (OCPP) allows a flexible connection of different charging stations to a backend system. It brings the advantage of being a de facto standard used by different providers. However, a connection requires the relevant partners to agree upon a common communication layer.

It was demonstrated that the selected roaming architecture works and is accepted by all the partners connected, as it supports current as well as future business models. The CROME partners recommend for a future marketplace to build a network of independent international partners (competitors) having their autonomous business and systems, the system design ensuring that each partner keeps his independence (data). It was proven, that real-time authorization between the partner systems linked via a roaming layer can be realized without any noticeable delay compared to the authorization within one partner system. This allows respecting the principle of minimized data storage and avoids large scale whitelist solutions.

Linking vehicle data to backend systems makes innovative services possible; however, specific legal framework conditions are to be considered. The roaming of services will be a key enabler for future mobility solutions.

Cross-border billing needs a clear legal framework

A clear legal framework is necessary for the implementation of cross-border billing, especially with respect to the billing of the value-added tax. Furthermore, showing a customized pricing table at the charging station of each provider may considerably increase the cost transparency for the end user. Similarly, transferring the pricing table via roaming service layer in connection with the authentication service also seems to make sense.

2 Data collection and storage

Daniel Ried

Institute of Applied Informatics and Formal Description Methods, Karlsruhe Institute of Technology

In order to analyse the usage of EV and their drivers' behaviour, large amounts of data were collected, stored and processed during the project runtime. For this reason a data repository was developed to ensure the secure, safe and reliable handling of any kind of electromobility data. The core component of this data repository consists of several relational databases for raw data, provided by EV drivers and manufacturers, and for processed and aggregated data (for further processing and analysis). The databases are surrounded by interfaces, tools, and services that were developed to provide for the import and the processing of raw data (including cleaning, harmonisation, and aggregation). The subsequent export of filtered and aggregated data and extensive analyses are based on a unified data model and commonly used data and file formats.

Before the development of the E-Mobility repository was started, various questions concerning data security and privacy issues as well as the data exchange and transfer process between the involved project partners had to be answered. These tasks were part of the planning phase, described in section 2.2. Their results were the foundation of the conceptual design of the repository, presented in section 2.3. They also served as the basis of the accompanying set-up of the technical infrastructure as well as the development of interfaces, tools, and services for data handling, described in section 2.4. Section 2.5 presents some selected experiences and challenging issues that occurred during the operation and maintenance of the data repository. Section 2.6 concludes this chapter with a summary and an outlook towards future scenarios.

2.1 Planning data collection

During the planning data collection phase the focus was on three main topics: (1) the compliance of the E-Mobility repository to the security and privacy requirements of the data owners, (2) the negotiation of a project-wide standardized data transfer process including manual and automatic actions with corresponding transfer protocols, and (3) the agreement of each data provider to a unified data exchange format including data schemas and file formats.

Most of the data collected contained confidential information owned by individual drivers, companies operating EV fleets, or manufacturers participating in the CROME fleet test. These confidential data are personal data like e. g. age, sex, profession, or income of drivers and fleet managers, person- or vehicle-related data like e. g. driving behaviour, locations, geographic coordinates, or technical vehicle data like e. g. energy consumption, range, or wear. In order to prevent any confidential data from being retraceable to persons and vehicles, they were pseudonymized by means of randomized driver and vehicle identifiers. Names, addresses, license plate numbers, and vehicle identification number (VIN) were not collected or stored. Furthermore, technical data are property of every single manufacturer and they must not be accessible to any other project partner. Thus, the hosting environment of the repository was set up within the KIT intranet infrastructure with multi-stage physical and logical access control and encrypted data transfer. This allowed on the one hand for the prevention from unauthorized access to the server location and to the network as well as the interception of messages and on the other hand for the protection of all data against abuse, corruption and manipulation.

Due to the heterogeneity of the individual sources of technical vehicle data a consistent data transfer process had to be defined in close cooperation with the EV manufacturers. The process was to be based on standardized interfaces and protocols as well as data and file formats. Furthermore, it had to comply with common security and privacy requirements, particularly with those of each EV manufacturer. For technical data the transfer process was defined as a recurring upload of text files. Technical data were collected throughout the project runtime by the EV manufacturers with logging devices embedded in certain customer EV. However, these devices were not equipped with wireless data communication technology such as the telecommunication standards Universal Mobile Telecommunications System (UMTS) or Long Term Evolution (LTE), so all collected data had to be read-out manually by the manufacturers along with regular maintenance work within fixed service intervals. Additionally, the collected data had to be automatically or manually preprocessed by the manufacturers in order to conform to their own requirements and to the specified data and file formats. The preprocessing included e. g. the pseudonymization of customer and vehicle identifiers, the filtering and aggregation of confidential but relevant data, the harmonisation of data values – more precisely their transformation from proprietary units and formats to the specified ones – as well as the assembling of the files prior to the upload. Figure 4 outlines the overall data transfer process for technical vehicle data.

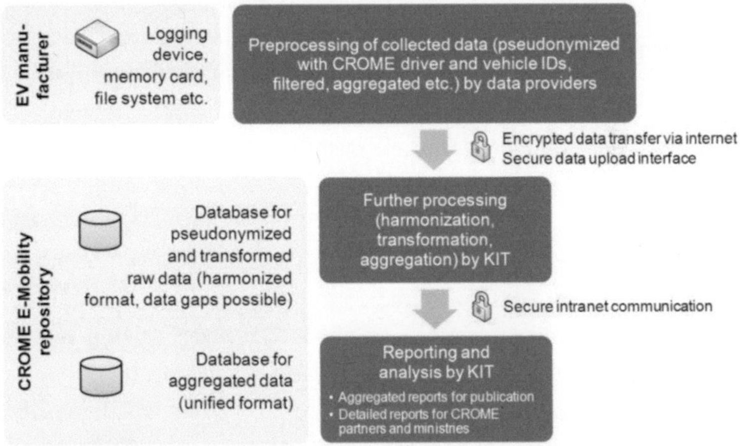

Figure 4: Data transfer process and processing of technical vehicle data for the CROME data repository at KIT.

As the upload intervals spanned several months, the number of files – and therefore also the data volume – was estimated to be very high. As a consequence, a file upload web site interface for large data volumes was developed that provided password-protected access to the E-Mobility repository and Secure Sockets Layer (SSL)-encrypted bulk file upload via the internet protocol Hypertext Transfer Protocol Secure (HTTPS).

Requirements specification for a harmonized data exchange format has been elaborated with regard to the data itself, its structure and values. This data exchange specification comprised those attributes which were identified as mandatory for the analysis of the EV usage and the drivers' experiences. Table 2 lists the required attributes (REQ) and Alternatively needed attributes (ALT) that most of the EV manufacturers were able to provide, e. g. the current timestamp, Global Positioning System (GPS) coordinates, temperatures, utilisation of electric consumers, mileage, remaining range, speed, pedal positions, state of charge (SOC) and power train data, as well as internal attributes (INT) used for the management and the traceability of the collected data.

Some of the EV manufacturers were not able to provide all of the above mentioned attributes; especially geographic coordinates could not be recorded due to technical limitations of their logging devices. Thus, smartphones were delivered to their customers (EV drivers or fleet managers) in order to collect geographic data and also other supporting data that had not been provided (cf. chapter 7). These data were recorded by the KIT mobility tracker application using the smartphones' position and motion sensors, e. g. GPS coordinates, altitude and speed. Additionally, other sensor data were collected to support further driving behaviour analyses, e. g. the acceleration and orientation of the smartphone for each coordinate axis. Table 3 lists those technical attributes which specify the data collected via smartphones including supporting data (SUP).

Table 2: Data exchange specification of the data repository (simplified).

No	Attribute	Unit	Type
REQ-1.1	DateTime	Universal Time Coordinated (UTC)/ISO 8601	Date and time value
REQ-9.1	GPS_Latitude	° or arcmin (arc minute)	Decimal value
REQ-9.2	GPS_Longitude	° or arcmin	Decimal value
REQ-10	GPS_Altitude	m	Decimal value
REQ-6	Temp_Out	°C	Decimal value
REQ-12.1	Temp_In	°C	Decimal value
REQ-15	Heating	on/off	Integer code
ALT-15.1	Heating_SeatLeft	on/off	Integer code
ALT-15.2	Heating_SeatRight	on/off	Integer code
REQ-16	AirCondition	on/off	Integer code
REQ-17	DrivingLight	on/off	Integer code
REQ-19	RemainingRange	km	Integer value
REQ-20.1	Speed	km/h	Decimal value
REQ-22.2	OperatingMode	driving/charging	Integer code
ALT-22.3	Ignition	on/off	Integer code
REQ-24	Mileage	km	Integer value
REQ-25	AccelerationPedal	%	Decimal value
ALT-26	BrakePedal	on/off	Integer code
REQ-28	SOC	%	Decimal value
REQ-44.1	Motor_rates per minute (RPM) (only for PHEV)	1/min	Decimal value
REQ-44.2	Motor_Torque (only for PHEV)	Nm	Decimal value
REQ-44.3	Generator_RPM	1/min	Decimal value
REQ-44.4	Generator_Torque	Nm	Decimal value
REQ-46.1	EnergyConsumption	kW	Integer value
REQ-49	Motor (only for PHEV)	on/off	Integer code
INT-1	Source	File name	Character value
INT-2	LfdNr		Integer value

Table 3: Geographical and technical data collected by the smartphone for the data repository (simplified).

No	Attribute	Unit	Type
REQ-1.1	DateTime	UTC/ISO 8601	Date and time value
REQ-1.2	GPS_DateTime	UTC/ISO 8601	Date and time value
REQ-9.1	GPS_Latitude	° or arcmin	Decimal value
REQ-9.2	GPS_Longitude	° or arcmin	Decimal value
REQ-10	GPS_Altitude	m	Decimal value
REQ-20.1	GPS_Speed	km/h	Decimal value
SUP-3.1	Accelerometer_X		Decimal value
SUP-3.2	Accelerometer_Y		Decimal value
SUP-3.3	Accelerometer_Z		Decimal value
SUP-4.1	Magnetic_Field_X		Decimal value
SUP-4.2	Magnetic_Field_Y		Decimal value
SUP-4.3	Magnetic_Field_Z		Decimal value
SUP-5.1	Orientation_X		Decimal value
SUP-5.2	Orientation_Y		Decimal value
SUP-5.3	Orientation_Z		Decimal value
INT-1	Source	File name	Character value
INT-2	LfdNr		Integer value

Unlike technical data from the vehicles, the data collected with smartphones could be transferred in real-time during the whole project runtime via internet using the smartphones' data connection. Hence, this data transfer process was also based on standardized interfaces and message protocols and the data and file formats were similar to those of the technical vehicle data (see above). Since the KIT mobility tracker application allowed for an automatic transfer of small text files containing the collected data, a web service interface was developed which could be automatically invoked at the end of each trip via the SSL-encrypted internet protocol HTTPS. The web service was implemented using XML-based standards like Web Services Description Language (WSDL)[4] and Simple

[4] WSDL is an XML-based language for describing the functionality of web services. It was developed and standardized by the World Wide Web Consortium (W3C); http://www.w3.org/TR/2007/REC-wsdl20-20070626

Object Access Protocol (SOAP)[5]. However, the access to the web service was restricted and only permitted to the delivered smartphones by means of an encoded username and password combination based on the smartphones' International Mobile Equipment Identity (IMEI) numbers[6].

For both, data provided by the EV manufacturers and data collected via smartphones, the sample ratio was set to ten data sets per second. Each data set was represented by a line of semicolon-separated values in a text file. Each file contained a chronological sequence of lines depicting a continuous and closed item of driving (a trip) or charging action. This sequence was preceded by a set of invariant data related to a single trip or charging process in terms of a file header. On the one hand such data were included in the text files uploaded by the EV manufacturers, and on the other hand they could be entered via the start screens of the KIT mobility tracker and afterwards sent to the web service interface together with the recorded data. Table 4 lists all trip-related attributes, e.g. the driver and vehicle ID, the purpose of the trip, the number of passengers and the weight of the carried luggage.

Finally, personal and behavioural data that were collected by means of online questionnaires had to be stored in the E-Mobility repository, too. The answers of the questionnaires were imported in the same structure and format in which they were stored and exported from the online survey tool. Personal data such as age, sex, profession, or income of drivers and fleet managers were used to enhance the pseudonymized driving behaviour profiles in order to categorize them into separate target groups (cf. chapter 8). Figure 5 gives a summarized overview of all types of data sources of the E-Mobility repository and of related import interfaces for the data transfer processes.

As the figure shows, technical data and the data collected with the delivered smartphones were imported into the E-Mobility repository using

5 SOAP is an XML-based message exchange format. It was developed and standardized by the W3C; http://www.w3.org/TR/2007/REC-soap12-part1-20070427

6 IMEI is a globally unique serial number for mobile phones.

encrypted and password-protected web interfaces. The import of the questionnaire responses was based on an Extendable Markup Language (XML) files export from the online survey.

Table 4: Trip-related data specification of data collected by the smartphone for the data repository (simplified).

No	Attribute	Unit/Values	Type
SUP-10	TripPurpose	private/commuting/business	Integer code
SUP-11	CrossBorder	yes/no	Integer code
SUP-12	Other means of transport (MOT)	none/bike/train/conventional vehicle/airplane/other	Integer code
REQ-3.1	Adults		Integer value
REQ-3.2	Children		Integer value
REQ-3.11	PassengerSeat	yes/no	Integer code
SUP-13	Luggage	5 kg/6-25 kg/26-50 kg/>50 kg	Integer code
REQ-4.1	User_ID		Character value
REQ-4.2	Car_ID		Character value
REQ-8	Car_Type		Character value
INT-3	IMEI		Integer value
INT-4	Trip_ID		Character value

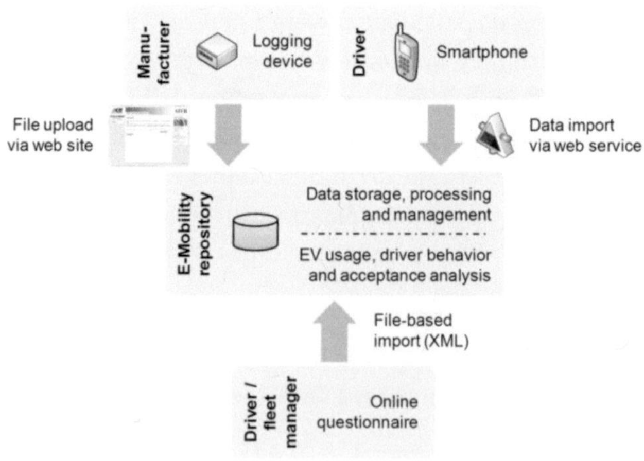

Figure 5: Data sources and data import interfaces of the CROME E-Mobility repository

2.2 Data storage design

The CROME E-Mobility repository consisted of several relational databases in order to persistently store raw data, which were provided by EV manufacturers and drivers, and aggregated data, which were the results of data processing such as cleaning, harmonisation, as well as of certain analyses. The data schemas were defined by means of flat Standard Query Language (SQL) database tables as a union of the set of trip-related attributes and, on the one hand, the set of technical vehicle attributes and, on the other hand, the set of smartphone attributes. As a consequence, sparse table entries and redundant data which led to an average overhead of at most 3 % of the total data volume, which was accepted to avoid additional join operations in queries later on. Additionally, the online questionnaires were stored in separate database tables conforming to the data schemas of the online survey tool.

In order to link all types of data for analytical purposes, i. e. technical vehicle data, personal and behavioural data, a set of reusable core business objects was identified and conceptually designed in close cooperation with the other involved KIT institutes. These objects were intended for the project-internal reuse of analysis results and they were also the foundation of a reference model for diverse and novel electromobility services. The reference model may enable a standardized access to data-centric basic services and an easy composition of complex value-added electromobility services. It may particularly allow for the processing and the representation of selected information for dedicated target groups and for various context-dependent purposes (cf. section 2.6). Business object classes like "User" (driver or fleet manager), "Car", "Trip", "Charging", and "ChargingStation" were encapsulated by specific interfaces and services that may provide generic queries with pre-defined search parameters, e.g. for offering personalized access to a chronological overview of an EV driver's trips (a history or a logbook), his range of movement or his energy consumption over a selectable period of time, or for visualizing anonymized information about all trips in a certain region. Figure 6 shows an excerpt of the unified data model consisting of core business object classes. It was modelled by means of a Unified Modelling Language (UML) class diagram and defined as a set of SQL database tables which were related to each other.

Each business object class represents a schema for a kind of fact sheet of related real world objects. According to this, a user's profile contains personal data based on the questionnaires' answers, modeled as attributes (within the class box), compositions with (lines with a black diamond) and associations to (named arrows) other classes. Such data reflect the user's role during the fleet test, his age, sex, education, home location, and optionally the company he was employed at. Further, it contains the user's driving and charging behaviour depicted by distributions over categories of covered distances, speed values, trip areas and SOC start values based on selective aggregations of all his trips and charging actions, respectively. Also the range of movement is an aggregated view of all the user's trips in the KML[7] format with an accumulated frequency of clustered GPS coordinates.

Trips and charging actions both associate a user and a car and they are both basically characterized by start and end values of date and time as well as SOC. The business object class for a trip consists of further attributes such as the covered distance, the average speed, the utilisation of heating and air condition, average inside and outside temperatures, the trip purpose, the number of passengers and the route as GPS coordinates in KML format. A charging action consists of a charging type (e.g. in private or public space) and a GPS location and it optionally associates a certain charging station.

Finally, further business object classes were conceptually designed for other relevant real world objects, i. e. for EV used in the CROME fleet test, for companies operating EV fleets, and for EVSE provided by project partners. The business object class for EV was separated into two subclasses for private and company vehicles, both based on the same superclass. The

7 KML is an XML-based language for describing and visualizing geographic data. It was developed by Google and standardized by the Open Geospatial Consortium (OGC); http://www.opengeospatial.org/standards/kml

superclass contains relevant static attributes describing the general product features of an EV corresponding to its manufacturer. Additional attributes describing the operational environment and surrounding conditions of a company car are put in its own subclass. Moreover, general information about a company is described by a distinct business object class. An EVSE is characterized by attributes for its address and GPS coordinates as well as for the technical specification of its charging spots. It also contains attributes referencing the provider and the manufacturer of an EVSE as well as other dynamic attributes characterising its state and behaviour, e. g. the availability and the reservability of single charging spots. However, some of these attributes are provided by the unified data model, but they are omitted at this point for legibility reasons.

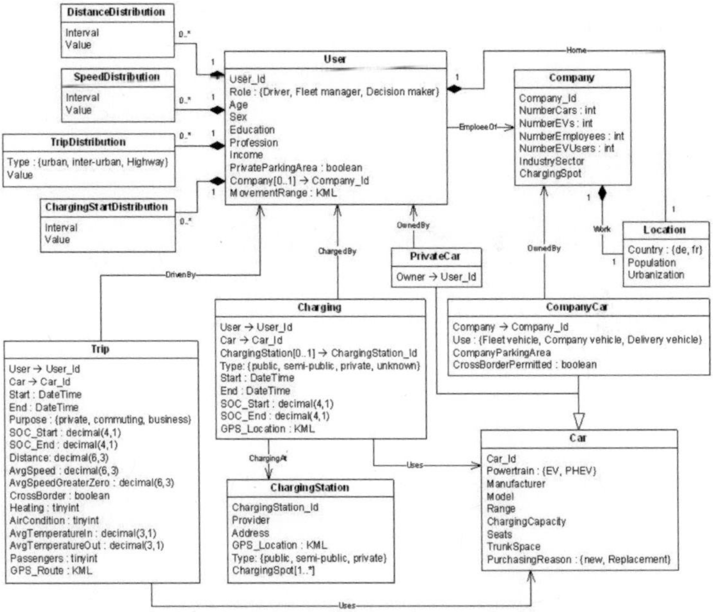

Figure 6: Unified data model for aggregated data of the CROME E-Mobility repository (excerpt).

2.3 Data repository development

After planning the data collection and conceptually designing the data storage, essential data handling routines had to be developed. This included, first of all, the implementation of the aforementioned software interfaces for importing raw data into the database conforming to the data exchange specification.

The upload web interface for large files was implemented as part of a web site which was provided separately to every single EV manufacturer (for security and privacy reasons). Each of the web sites included a manufacturer-specific authentication and authorization module as well as web pages for the management of user accounts and privileges, e.g. for creating user accounts for the people in charge of assembling and uploading files and changing a user's password. However, the main part of each web site was the data import module that was invoked each time a file was uploaded by an authorized user. Figure 7 shows the web page form for uploading files.

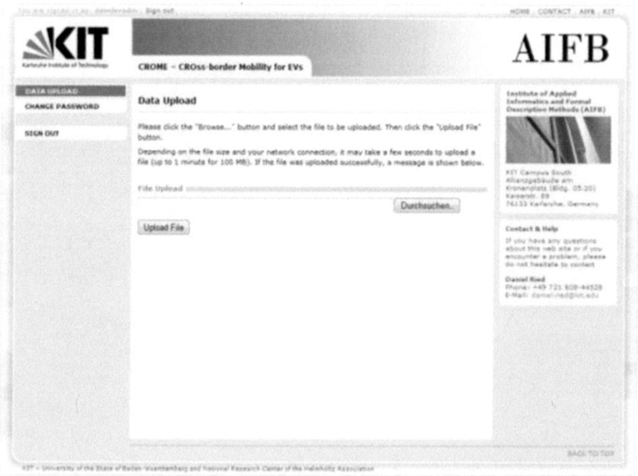

Figure 7: File upload web site of the CROME E-Mobility repository for EV manufacturers.

The uploaded files were temporarily saved on the file system of the server hosting the web sites and were automatically processed and imported into the databases. The processing was done in blocks of several files with a total size of 100 MB at a scheduled point of time in the evening of the same day. While processing the files, their text-based contents were checked for conformance with the data exchange specification, i.e. the correctness and consistency of the file format and the data structure (file header and payload data).

The processing and import routines had to be continuously adapted and improved during the software development and the testing because the data provided by the manufacturers were not entirely of the quality expected. The main issues discovered during test imports in an early stage of the project were about attribute values that were recorded in an alternative proprietary unit (e. g. arc minute instead of degree used as unit of GPS coordinates) or scale, missing values (so-called NULL values), different characters used as decimal points (point or comma) or value separators (semicolon or comma), and the ambiguousness of date and time values. All these data anomalies had to be kept in mind during the software development and finally they were automatically handled by the application logic of the data import module.

The web service interface was developed for the automatic real-time transfer of small data volumes. A web service is usually provided by a software component that is invokable over the internet by other software components or applications, such as the KIT mobility tracker smartphone application. However, the data upload web service could only be accessed by this particular application installed on the delivered smartphones. The size of the uploaded files was implicitly limited by the KIT mobility tracker to approximately 2 MB, since the data were transferred as trip fragments, each spanning exactly 20 minutes. All files, which were uploaded by any of the smartphones, were temporarily saved at the same place on the web server's file system and they were automatically imported into the databases with a scheduled delay of five minutes after the upload of the last file. The data exchange specification was accurately implemented by the

KIT mobility tracker, so no attribute value conversion had to be done during the import routine, but the date and time attribute values had to be checked for duplicates followed by the elimination of obsolete data sets. This was due to an inaccuracy while recording data from the position and motion sensors in short intervals, which results in repetitive data sets with identical timestamps.

The export of the data was realized on a regular basis for all types of data. For this purpose a harmonized semicolon-separated export file format was specified together with the involved KIT institutes. Each of the exported files consisted of a filtered set of attributes required for potentially significant analysis results. Each file contained a header of invariant data and a chronological sequence of contiguous data sets, where the time difference between each two consecutive datasets was less than or equal to five minutes. Otherwise, if the difference from one data set to its preceding one was greater than five minutes, a new export file was created. Unlike with trips, the export files for charging actions included not all available data sets, but the number of data sets was reduced to one per five seconds. The file-based data export was supported by a software tool which was developed and refined during the early stages of the project. It provided parameterized queries, e. g. for the export of available trips and charging actions of certain drivers over a selectable period of time.

The export of recently imported data was performed manually every few weeks using the export tool. During the export, all files were checked for consistency and quality and the thereby identified deficiencies had to be corrected. Thus, for instance, short trips with too less data sets were eliminated; such trips were generally recorded by mistake, e. g. when (dis)connecting the EV to (from) the EVSE, or when the mobility tracker application was started unintentionally. All remaining files were of high quality and thus seemed to be a convenient basis for the following EV usage and driving behaviour analyses (cf. chapter seven). Another deficiency, which was identified during the analyses later on, was about mostly large export files, which contained a huge sequence of consecutive but partly obsolete data sets; such data sets were recorded although the EV was not operated for a certain period of time, e. g. when the mobility tracker was not stopped directly after a trip had been finished. The method for detecting obsolete data fragments and for separating single files into several trips using the mathematical computing environment Matlab is described in detail in chapter seven. Finally, a semi-automatic but still effort-consuming iterative data cleaning process was carried out continuously in order to detect all obsolete data sets and delete them from the databases by means of specific executable SQL scripts, which had to be generated from results of the aforementioned trip separation method. Figure 8 shows the summarized overview of all types of export interfaces of the CROME E-Mobility repository, which were developed during the project runtime and which have been provided to the involved KIT institutes.

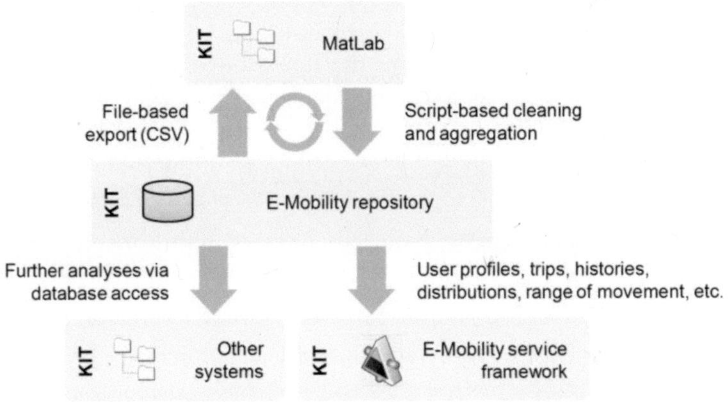

Figure 8: Iterative data processing and export interfaces of the CROME E-Mobility repository.

The E-Mobility repository in the center of figure eight is equipped with interfaces to the surrounding applications, systems or frameworks. For the processing of the data and their extensive analyses, the data were provided via a file-based export conforming to the business object classes of the unified data model using widely accepted file formats, such as comma- or semicolon- separated values (CSV) as well as other formats like e. g. XML and JavaScript Object Notation (JSON). Alternatively provided access to the data repository was realized on the one hand by a direct database access for other analysis or statistics tools, and on the other hand by the concept of the E-Mobility service framework based on web service interfaces.

2.4 Operation and maintenance

The E-Mobility repository had been developed and set up since May 2011. The data collection lasted from November 2011 until October 2014. About 564 million raw data sets were imported into the repository during this period. This number was reduced to 432 million data sets after cleaning and trip separation. The total size of all databases was added up to nearly 150 gigabytes, which had to be managed and maintained during the whole project runtime, e. g. all databases had to be backuped automatically on a weekly basis; additionally, the previously uploaded files were archived after their import. Finally, the volume of the filtered and aggregated data was about 22 gigabytes, divided into 120,000 trips and 6,800 charging actions.

Besides operating and maintaining the databases of the E-Mobility repository there were challenging issues of many kinds that had to be monitored and coped with. These were about the servers' hardware, the network topology, the servers' operating systems, the repository's software interfaces, tools, and services, and, in particular, general security, safety and reliability issues of all hardware and software components.

2.5 Conclusion and outlook

In order to be able to analyse the EV usage and the EV driver behaviour, it was necessary to collect large amounts of data during the fleet test. For this purpose, an E-Mobility repository was developed compliant to the security and privacy requirements of the EV manufacturers and drivers. The core components of the E-Mobility repository were several databases for the storage and the management of heterogeneous data sets collected and transferred by the EV drivers and manufacturers, who participated in the CROME project. Additionally, various interfaces, tools, and services were developed which provided for the import and processing of raw data as well as for querying, accessing and exporting aggregated business objects for further analytical purposes. Such business objects conform to a unified data model which was also the conceptual foundation of a reference model for an E-Mobility service framework. The latter consists of specific interfaces and services which may enable a standardized access to data-centric basic services and an easy composition of novel value-added electromobility services.

In future use cases this reference model may allow for the processing and the representation of selected information for dedicated target groups, e. g. drivers, fleet managers, EV manufacturers, EVSE operators, utilities, municipalities or other third party service providers. Possible integration scenarios of services based on the reference model may be (mobile) web applications, smartphone applications, or applications integrated in an in-vehicle infotainment system. These applications may offer text- and table-based representations including customizable queries and filter functions as well as graph- and map-based visualizations, e. g. by means of Google Maps or Open Street Map.

As an exemplary use case, a composite service may provide information that is not available so far, but may be potentially relevant for EVSE operators, utilities or municipalities. This service can be composed in order to link an anonymized aggregation of all trips in a certain region with the positions of nearby EVSE or any other data provided in the same way by other parties, like e. g. traffic information, positions of traffic lights, or infrastructure capacity. Figure 9 shows an aggregated and clustered visualization of all trips in 2013 in the region of Offenburg/Lahr as a heat map and available EVSE located there as marker points by means of Google Maps.

On the heat map visualization, road sections with a high volume of EV traffic are highlighted in red. This may provide the aforementioned target groups with an added value, e. g. when planning further development of the charging infrastructure and, in particular, the construction of sections for dynamic inductive charging lines. This may also be an additional aspect of dynamic traffic management in the future, e. g. when planning the installation of fast or bus lanes, on which EV are permitted to drive (during certain hours or traffic situations).

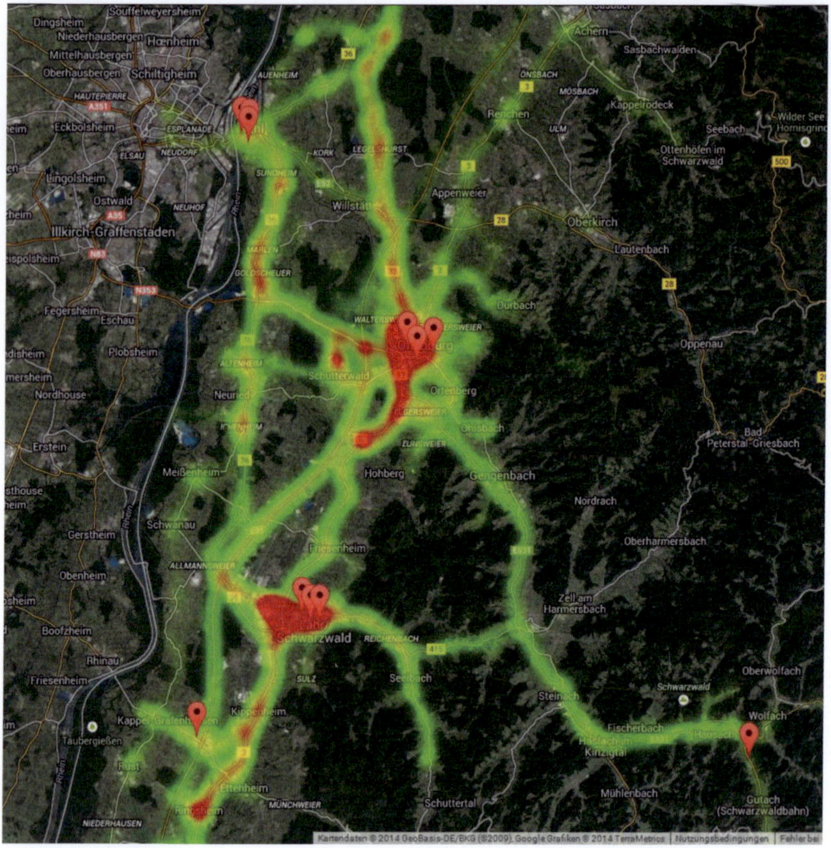

Figure 9: Aggregated visualization trips and EVSE of the CROME fleet test based on Google Maps.

3 Optimized strategies for DC-charger grid access

Peter Krasselt

Institute of Electric Power Systems and High-Voltage Technology, Karlsruhe Institute of Technology

The charging requirements imposed by the majority of EV trips can be satisfied by AC charging with connection powers from 3.7 kW up to 22 kW (NPE 2014). This may be easily accomplished in compliance with existing technology. However, for long distance trips or flexible spontaneous route planning, a fast charging infrastructure is necessary.

In 2020, a need for a public charging infrastructure consisting of 173,000 normal power recharging points (22 kW, AC charging) and 7,100 high power recharging points (50 kW DC charging) is predicted in Germany (NPE 2014).

Connecting fast-charging DC-EVSE to the LV electricity network has a significant impact on power quality.

3.1 Standards and technical rules for EVSE network connection

In order to ensure the standardized voltage characteristics in public distribution networks (CENELEC 2010), it is mandatory to comply with a number of relevant standards, technical rules and regulations when connecting a DC-EVSE in Germany. As demanded in (BDEW 2011), the EVSE operator requests a network connection for a load exceeding 12 kW rated power from the distribution network operator, since the common range of standardization for loads connected to low-voltage networks is limited to currents up to 75 A (DIN 2001, DIN 2012) and typical DC chargers are grid connected with a nominal current of at least 80 A. An individual technical assessment of network disturbance (D-A-CH-CZ working group EMC 2007) has to be done. In this chapter, the following aspects are considered:

First, the characteristics of EVSEs with passive and active front-end rectifiers are summarized. Next, typical LV network characteristics are analysed, as the degree of network disturbance has to be calculated at the point of common coupling (PCC).

This is followed by a discussion of the most relevant disturbance aspects. Finally, an optimized DC-EVSE AC mains connection is proposed.

3.1.1 EVSE operation characteristics

The overall efficiency η of a DC-EVSE, measured from the AC power entry to the DC charging connector, can be estimated as 95 %. With a DC charging power of 50 kW, the rated AC power at unity power factor is 53 kVA, resulting in an AC three phase power connection request of at least 3x80 A.

The key component for the impact on power quality in LV-networks is the rectifier topology. Most EVSE interface AC mains with a passive line-commutated rectifier as shown in Figure 10a. The passive rectifier type draws an inductive power factor λ of 0.96 at rated power operation. The total harmonic distortion (THD_I) of the line currents is limited to the range of 40 to 70 % by technical measures (AC commutating reactor L_{AC} and DC smoothing reactance L_{DC}).

The EVSE in Figure 10b is equipped with an active front-end. Using IGBTs, this self-commutated rectifier draws a nearly sinusoidal current with a THD_I lower than 10 %. This rectifier topology also features full independent control of the reactive power consumption.

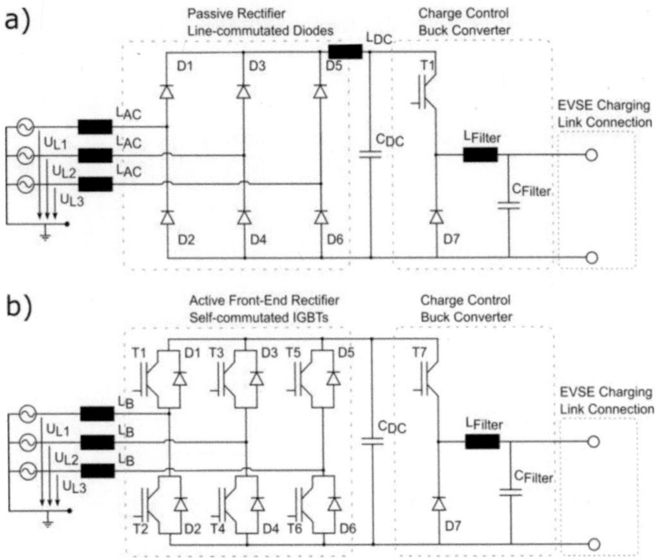

Figure 10: Considered EVSE-Topologies.
a) standard line-commutated rectifier
b) proposed self-controlled active front-end rectifier

3.1.2 Network characteristics

A key factor for the assessment of network disturbances is the grid impedance at the PCC of the EVSE. The impedance at a PCC can be calculated by using the simplified network diagram shown in Figure 11.

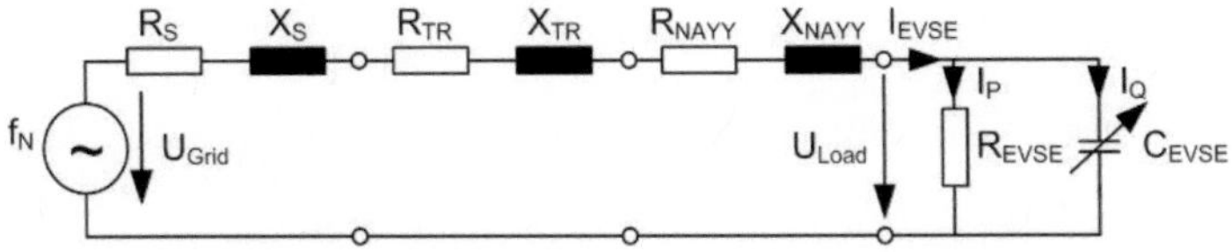

Figure 11: Single-line diagram of LV-network with an EVSE.

The high-voltage and medium voltage network are modeled in the source resistance (R_S) and impedance (X_S). A standard local substation 20 kV/0,4 kV transformer with a rated power of 400 kVA feeds a LV cable type NAYY4x150mm^2. This is the most common cable type in German LV networks (Kerber 2011). With the parameters given in Table 5, the grid impedance is expressed using the short-circuit power at the PCC, and the R/X ratio is calculated for variable cable lengths (Figure 12).

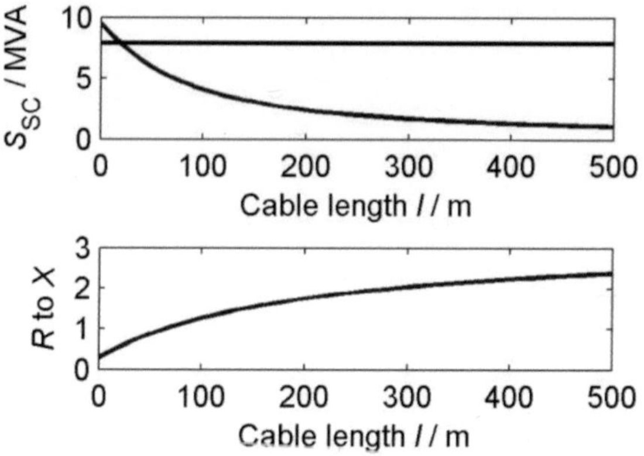

Figure 12: Main grid impedance parameters for a low voltage grid.
Upper plot: Short-circuit power (constant), required short-circuit power for uncontrolled rectifier.
Lower plot: R/X ratio for variable LV cable length

Table 5: Network parameters for grid impedance evaluation.

Parameter	Value	Unit	Description
U_{Grid}	400	V	Nominal voltage
R_S	620	µΩ	Source resistance
L_S	2.6	µH	Source inductance
f_N	50	Hz	Power frequency
R_{TR}	5.7	mΩ	Transformer resistance
L_{TR}	70	µH	Transformer inductance
R'_{NAYY}	256	mΩ/km	Resistance per unit length
L'_{NAYY}	255	µH/km	Impedance per unit length

3.1.3 Assessment of network disturbance

Harmonic voltages

Rectifier circuits inject harmonic currents in the power system. Due to the line impedance, harmonic voltages build up. In case of an EVSE with a 6-pulse *line-commutated rectifier* (Figure 10a), a short-circuit power 150 times higher than EVSE rated power (shown for a 50 kW charger as black trace in the chart, Figure 12) is required for grid access (D-A-CH-CZ working group EMC 2007). The critical line length providing this amount of short-circuit power is 20 m. For longer cable distances, additional measures have to be taken for harmonic current limitation, e.g. use of a 12-pulse rectifier fed by a three-winding transformer.

An EVSE featuring a self-commutated rectifier meets harmonic emission requirements with ease. With a THD_I smaller than 10 %, no harmonic assessment has to be done. To suppress interharmonic voltage emission, the IGBT switching frequency has to be locked to an integer ratio synchronous to the grid frequency.

Rapid voltage changes

Fast fluctuations in electrical power consumption generate rapid voltage changes. While starting a charging process, EVSEs avoid flicker emission with a slow power-on ramp over several seconds. During charging and towards the end of the charging process, power fluctuation are small, so rapid voltage changes are no limiting factor for the installation of EVSE.

Supply voltage variations

The charging time of an EV is estimated to be at least 10 minutes. Therefore, the change rate r for supply voltage variation is assessed to r = 0.1 min-1 and the maximum tolerated voltage drop $d_{max,LV}$ due to charging activity is 3 % of nominal network voltage U_{Grid} (D-A-CH-CZ working group EMC 2007). This crucial criterion defines the maximal line length between a local substation and an EVSE and is investigated in detail in this study.

3.1.4 Proposed EVSE topology

Offering the feature of independent regulated reactive power and low harmonic current emission, an EVSE with a self-commutated active front-end is selected for optimized grid integration. The active-rectifier is designed for a power factor of 0.9 at rated power, resulting in a total connection power of 60 kVA fed by a three phase AC connection 3x100 A.

3.2 Reactive voltage support

3.2.1 Expansion of maximal line length

Analytical approach

First, the principle effect on supply voltage variations during EVSE operation and the reactive power support to reduce voltage drop is analytically investigated. The voltage drop U_Δ during EVSE operation at the EVSE PCC is expressed in equation 1 by applying the Kapp triangle shown in Figure 13.

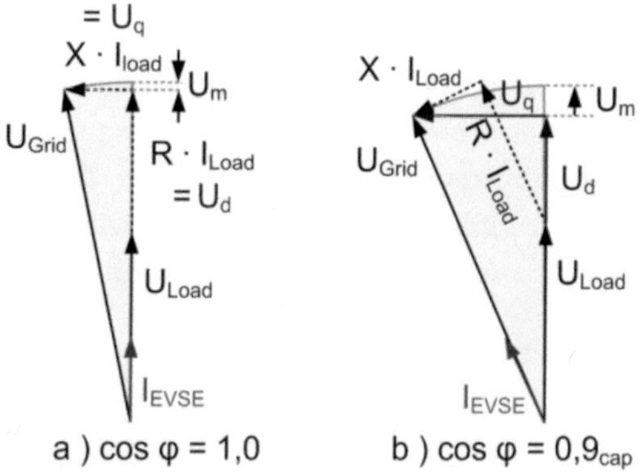

Figure 13: Kapp triangle for a) unity and b) capacitive-resistive power factor.

$$U_{\Delta} = U_{Grid} - U_{Load} = U_d + U_m \tag{1}$$

$$U_d = R \cdot I \cdot \cos\varphi + X \cdot I \cdot \sin\varphi \tag{2}$$

$$U_q = X \cdot I \cdot \cos\varphi + R \cdot I \cdot \sin\varphi \tag{3}$$

Using the example LV network shown in Figure 12 the resistance R and inductance X is given by equations 4a to 4d,

$$X = X_S + X_{TR} + X_{NAYY}, \quad X_{NAYY} = X'_{NAYY} \cdot l \qquad \text{(4a), (4b)}$$

$$R = R_S + R_{TR} + R_{NAYY}, \quad R_{NAYY} = R'_{NAYY} \cdot l \qquad \text{(4c), (4d)}$$

where l is the cable line length. The current I is calculated from the EVSE apparent power drawn from the AC mains. The EVSE power factor is calculated by distinguishing its real current I_P and reactive current I_Q.

$$I = \sqrt{I_P^2 + I_Q^2}$$
$$I_Q = I_P \cdot \tan\left(\cos^{-1}\varphi\right) \tag{5}$$

Using the geometric relations defined in Figure 13, U_m can be defined by the following equation:

$$\left(U_{Grid} - U_m\right)^2 + U_q^2 = U_{Grid}^2 \tag{6}$$

By substituting U_m in (1) using (6), the voltage amplitude drop between U_{Grid} and U_{EVSE} is now defined by

$$U_{\Delta} = U_d + U_{Grid} - \sqrt{U_{Grid}^2 - U_q^2} \tag{7}$$

Calculation of maximal line length

The maximal cable length between the local substation and the PCC can be calculated by solving equation (1) for

$$U_{\Delta} = d_{\max,\mathrm{LV}} = 3\% \cdot U_{Grid} \tag{8}$$

calculating with the rated current value and unity power factor in (2), (3) and (5):

$$I_P = 80\ \mathrm{A} \quad \text{and} \quad \varphi = 0° \tag{9}$$

As we can see in the upper plot of Figure 14, the maximal line length is determined as 330 m. For longer distances, a capacitive power factor has to be increased to preserve the $d_{\max,LV}$ criterion. With respect to the minimal required power factor of 0.9 (BDEW 2011) for LV connected loads, the maximum line length can be increased by 30 % to 429 m (lower plot in Figure 14).

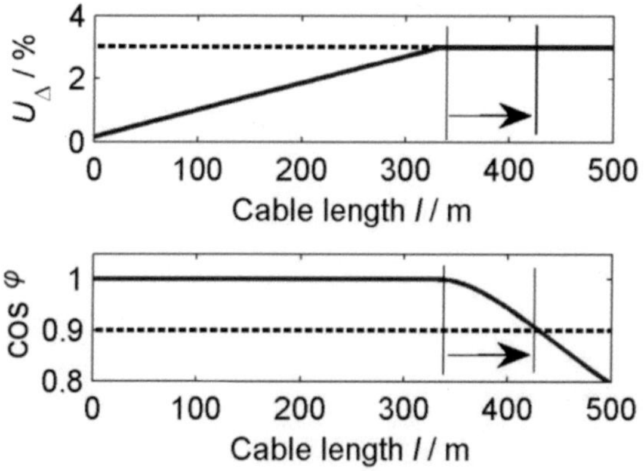

Figure 14: Derivating the maximal line length for the considered low voltage grid.
Upper plot: Voltage drop U_{Δ} for variable cable length l
Lower plot: Required cos φ_{EVSE} to ensure d_{max} = 3 %

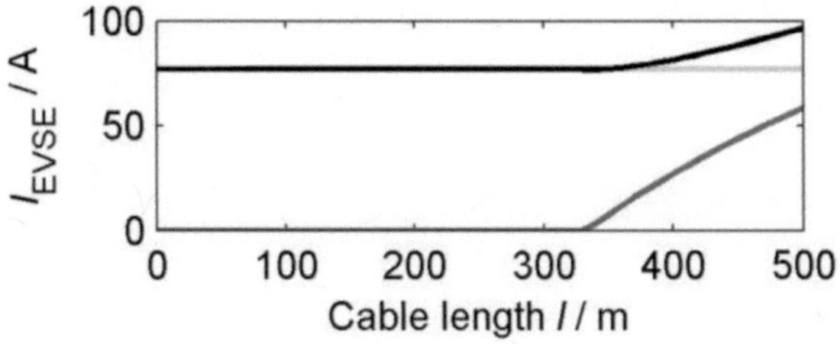

Figure 15: EVSE currents IEVSE, Rated current (light grey), active current I_P (black) and reactive current I_Q (dark grey).

The benefit of simplified grid access by reactive power support of EVSE comes at the cost of increased grid and inverter losses. With increasing reactive power consumption, the EVSE draws an increasing line current I_{EVSE} (cf. Figure 15).

With the increasing number of power converters for distributed renewable energy (DRE) generation (e.g. for photovoltaic systems) and the integration of new load types like DC-EVSE, the attributes of the reactive power control scheme should be defined carefully. Therefore, the efficiency of different approaches is discussed in the following chapter.

3.2.2 Detailed analysis of reactive power voltage control schemes

Methodology

When analysing the effects of different reactive power control schemes, the characteristics of power converters and distribution networks have to be considered. The voltage varies within the tolerance band during the course of the day. In addition, weekdays as well as weekends have to be distinguished, and network load is also influenced by seasonal effects. This requires analysis periods of 24 h, and an examination of the impact of different day characteristic and seasons. In addition, different network topologies have to be analysed.

On the other hand, power converter dynamics have to be modeled for loss estimation and correct step-response behaviour of the system. Power converters operate at switching frequencies of several kilohertz, requiring simulation time steps in the sub-microsecond range. Feasible simulation length is therefore limited to a range of some minutes.

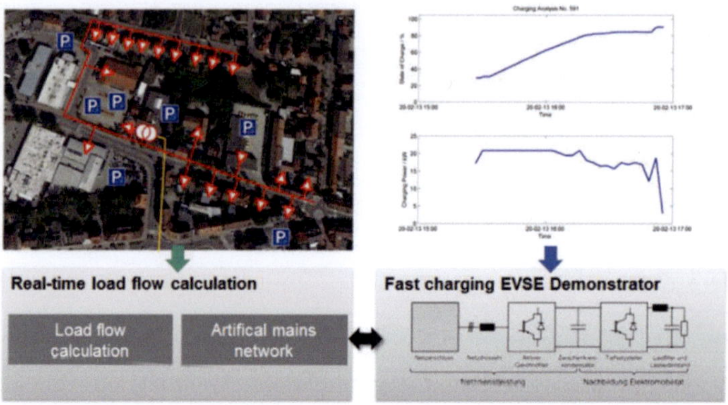

Figure 16: Outline of the developed hybrid approach for voltage-stability analysis and reactive voltage support scheme comparison.

In this investigation, a hybrid approach has been chosen to cover both correct power converter analysis as well as different grid scenarios. An experimental full-scale power converter emulating fast charging DC-EVSE operation has been built in hardware and connected to an artificial mains simulator as shown in Figure 16. This Grid-in-the-Loop configuration offers several important benefits:

- The artificial mains simulation guarantees identical, repeatable and freely definable network characteristics for the comparison of reactive voltage support schemes in different network environments.
- For pre-testing of network scenarios and power converter algorithms, the 24 h test scenarios could be accelerated for faster result validation.

- The experimental full-scale power converter could be conFigured to emulate any synthetic or real-world measured charging curve.
- As a real power converter is used, validity of power converter step-response behaviour and system losses is ensured for this particular hardware implementation.

Full-scale DC-EVSE emulation system

Topology

In section 3.1, technical requirements for an optimized grid integration of a DC-EVSE have been derived. Considering these requirements, a self-commutated active front-end for rectification has been proposed. Self-commutated rectifiers provide the advantage of drawing nearly sinusoidal line currents at unity power factor. In section 3.2.1, benefits of reactive voltage support for grid voltage stabilization have been analyzed. Therefore, an active front-end rectifier topology which can operate various power factors has been selected to feed the voltage link. To emulate DC-charging operation, a buck converter consumes energy from the voltage link, which is dissipated in a power resistor. The topology is shown in Figure 17 and technical specifications are shown in Table 6.

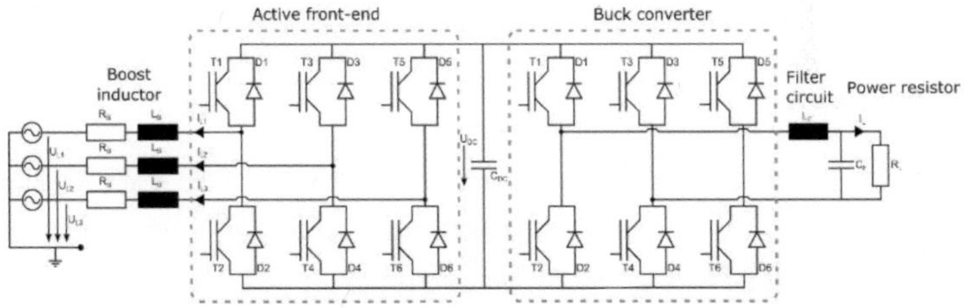

Figure 17: Power converter topology of the full-scale DC-EVSE simulation system.

Table 6: Technical specifications of the full-scale DC-EVSE emulation system.

Active front-end rectifier			
Rated power	S_N	[kVA]	80
Nominal AC voltage	U_N	[V]	400
Nominal AC line current	I_N	[A]	200
Boost inductator	L_B	[µH]	500
DC voltage link capacitance	C_{DC}	[mF]	1,0
Nominal DC link voltage	U_{DC}	[V]	700
Switching frequency	f_{AFE}	[Hz]	8,050
Buck converter emulating DC-EVSE operation			
Rated active power	P_N	[kW]	60
Nominal charging voltage	U_N	[V]	500
Nominal charging current	I_N	[A]	120
Power resistor	R_L	[Ω]	4,2
Filter inductor	L_F	[µH]	400
Filter capacitance	C_F	[µF]	200
Switching frequency	f_{BC}	[Hz]	8,050

To simulate DC-charging activity, a buck converter draws energy from the voltage link. An integral control regulates the power dissipated in the power resistor according to a reference signal $P_{DC,ref}$. The source of the reference signal is a set of values stored in selectable data files. The values have been preassigned by either computer simulation or by measurement and recording of real charging characteristics.

Control concept

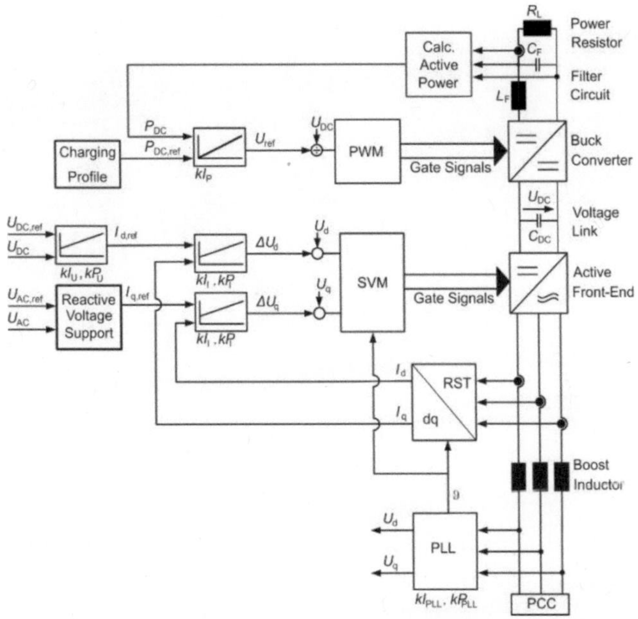

Figure 18: Power converter control concept of the DC-EVSE simulation system.

Artificial mains network simulation

Grid-in-the-loop concept

The *Grid-in-the-Loop* concept was developed to investigate the effects of a fast charging system on different grid structures. It connects a simulated virtual grid via an artificial mains to a power converter emulating a DC-EVSE system as introduced in section 3.1.4.

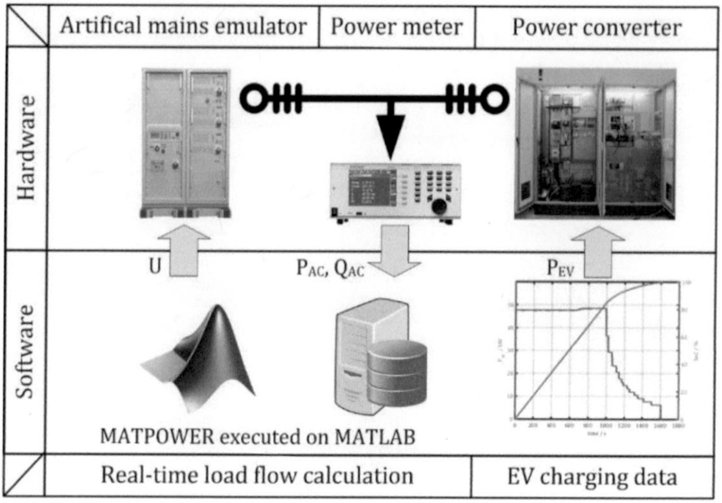

Figure 19: The Grid-in-the-Loop concept.

The *Grid-in-the-Loop*, represented in Figure 19, consists of four basic elements. The main part - the Matlab based real-time network power flow solver - includes the possibility to analyse different grid structures during a day. The four-quadrant converter is used as an artificial mains simulator, connecting the computer simulation of a LV-network and the emulated DC-EVSE system. The actual active and reactive power drawn from the network simulator is measured by a power meter and is taken into account for the next time step of the power flow analysis.

The *Grid-in-the-Loop* concept uses the power flow simulation package MATPOWER (Zimmermann et al. 2010) to evaluate network voltages and load flow on transmission lines. For a 24 h power flow analysis, the calculation is time-discretized with a step size of 1 s and the algorithm computes the analysis for each of those steps in a row.

All four elements of the *Grid-in-the-Loop* concept require some time to process the data. The amount of time for each component involved has been determined as shown in Figure 20.

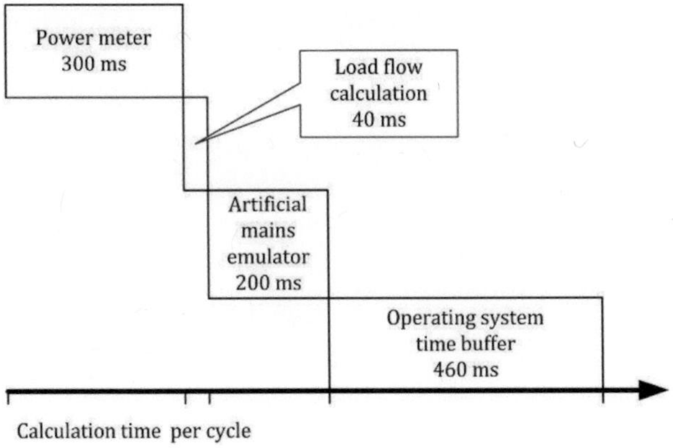

Figure 20: Timeline of a single calculation step of Grid-in-the-Loop concept.

A buffer for operating system delays of nearly 500 ms is necessary to ensure equidistant time-steps, limiting the maximum update rate to 1 Hz. As power flow variations are quasi-stationary and no grid fault dynamics are considered in this investigation, this update rate is sufficient.

The transmission lines and transformers are modeled with a common branch model. In addition to the standard grid elements, various types of load or generator profiles can be adapted. The consumption of domestic loads has been simulated using a behaviour based load profile which has been generated with the tool *Load Profile Generator* of the TU Chemnitz (Pflugradt et al. 2013).

The impact of DRE generation on voltage stability is assessed using a photovoltaic generator model. Power fed in the LV-network is modeled with respect to the geographic position and photovoltaic installation details (E. C. Institute for Environment and Sustainability 2014).

Reference scenarios

For the purpose of power flow analysis, reference grids have been introduced to assess the impact of adding DC-EVSE to existing LV-networks: Based on statistical data, the characteristics of the reference grids are typical for typical LV-networks in Germany (Kerber, 2011) (Scheffler, 2002). The requirements for modeling are:

- rated apparent power of local substations
- general network topology
- quantity and length of transmission lines
- photovoltaic generators (number and locations)
- demographic structure of the region, and
- total number of domestic loads.

The modeled reference grids provide the base to assess the voltage stability and network losses as a result of connecting DC-EVSE to the grid. To consider the typical characteristics of different grids – taking the photovoltaic generation and population structure in account – three distinguished reference scenarios have been selected: the *Village*, the *Suburb* and the *City* scenario (Table 7). All scenarios have in common that domestic customers are prevailing. However, the customers' demographic structure varies between the particular scenarios.

The vicinity of Karlsruhe served as reference to model the residential areas (Statistical Office of the Federal State of Baden-Wuerttemberg 2014). Cables of NAYY4x150mm^2 serve to model the transmission lines, as this cable type is most commonly found in German LV-networks. The installed photovoltaic generators are evenly distributed over the network.

Table 7: Reference scenarios for the LV network analysis: Village, Suburb and City.

PV-Generation	Venue	EVSE PCC	Substation	Grid structure	Reference Scenario
180 kW	Malsch	Cable segment 1, 315 m	240 kVA, u_k = 4 %	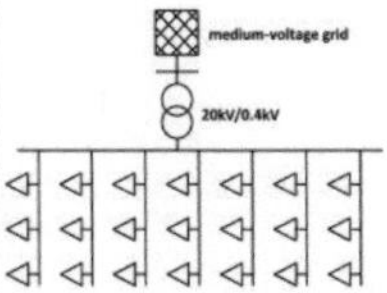	a) Village
150 kW	Karlsruhe Hagsfeld	Cable segment 4, 420 m	400 kVA, u_k = 4 %	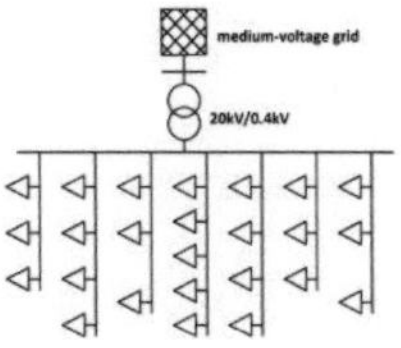	b) Suburb
130 kW	Karlsruhe Südstadt	Cable segment 2, 240 m	630 kVA, u_k = 4 %	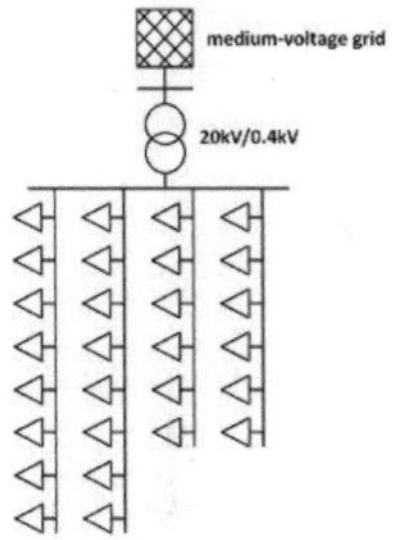	c) City

Table 8: Reactive voltage support control schemes analysed.

Power Factor Scheme Power converter type	Distributed renewable energy inverter	Adaption for high-power EVSE rectifiers
a) Fixed cos(φ)	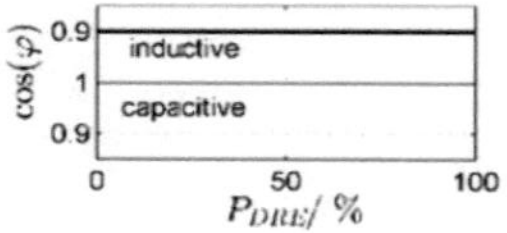 	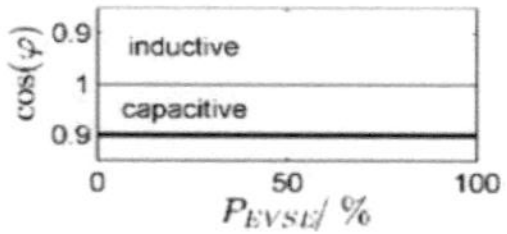
b) Active power regulatedcos(φ(P))	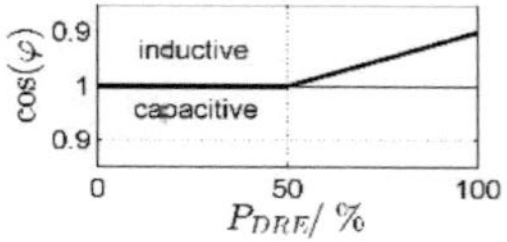 	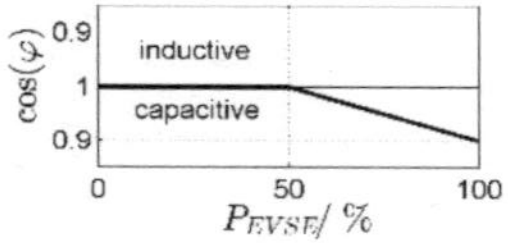
c) Voltage amplitude regu-lated cos(φ(U))	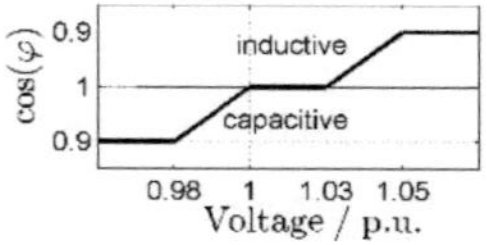 	
d) Hysteresis controlled cos(φ)	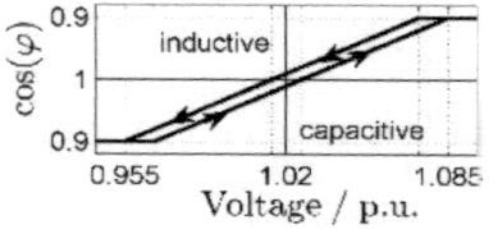 	

Reactive power voltage support schemes

On the foundation of (VDE 2011, Westnetz GmbH 2013), four different power factor schemes (8) have been implemented for further analysis:

The *fixed* and *active power regulated schemes* (8a and b) are mostly used for photovoltaic inverter systems directly connected to the LV-network. Providing inductive reactive power based on the instantaneous active power infeed, voltage boost is reduced to ensure voltage stability. In LV-networks with DRE and EVSE systems, this control scheme suffers the problem of conflicting policies: DRE systems provide inductive, EVSE capacitive reactive power, resulting in a partial reactive power cancellation.

The *voltage amplitude regulated* and *hysteresis controlled* reactive power voltage support scheme (8c and d) provide a solution to the problem of partial reactive power cancellation. Grid voltage measured at the PCC defines the required power factor in these schemes. Major benefits of the voltage-based power factor scheme are first a unique control method for both DREs and DC-EVSEs and second attenuating the phenomena of partial reactive cancellation. As network voltage varies over the cable length, each power converter operates at an individual set point controlled by its actual terminal voltage.

3.2.3 Results

Voltage stability

In a first step, all reference scenarios have been verified using the *Grid-in-the-Loop* concept. The power converter emulating DC-EVSE charging operates with one of the reactive voltage support schemes according to Table 8. An instance of a test trial, configured with a fixed power factor of 0.9_{cap}, is shown in Figure 21. Over 24 hours, six charging operations of different power levels are initiated by electric vehicles. It is shown that every time the injection of reactive power boosts the voltage and therefore ensures voltage stability.

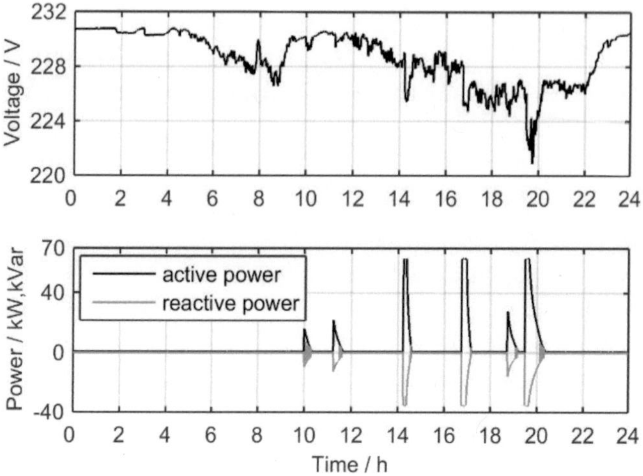

Figure 21: Measured power and network voltage at the PCC.

As analytically shown in section 3.2.1, capacitive reactive power provides the capability to increase the maximum possible line length. To confirm this result in more complex and advanced LV-networks, all reference scenarios were emulated with the DC-EVSE system working in different control modes. The procedures have been exercised separately for no DRE infeed (winter season) and an elevated level of DRE infeed (summer season).

All power factor schemes induce a voltage boost of roughly 0.5 % with respect to the nominal voltage (Figure 22) and therefore confirm the analytic results. This enables distribution network operators to permit the installation of DC-EVSE systems also at locations which are more distant from the local substation.

With increasing reactive power consumption, the EVSE draws an increasing line current. As higher line currents result in higher network losses, the impact of reactive power voltage support is examined using a variation of connected load parameters.

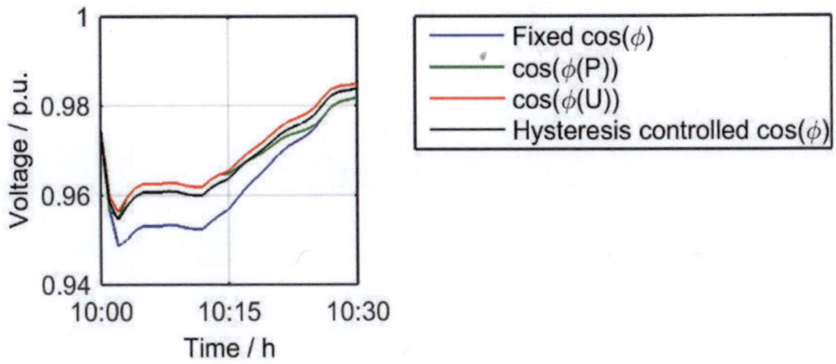

Figure 22: Voltage curve at connection point of the DC-EVSE, emulated using the rural reference scenario.

Evaluated scenarios for active power loss assessment

For analysing the effects of controlled EVSE systems on power losses in LV-networks, the three scenarios village, suburban and city have been simulated with different load situations. This accommodates the variety of LV-networks. For the instance of the city scenario and individually for the summer and winter seasons, the interactions of voltage support schemes of DRE and DC-EVSE are evaluated in the following. An instant of the losses for the different power factor schemes (cf. Table 8) are shown for the city scenario and compared to EVSE operation with unity power factor (Figure 23).

Without photovoltaic infeed, the hysteresis controlled power factor scheme causes the least additional losses compared to a unity power factor. The losses are about 4 % lower than using the other schemes and the hysteresis control exhibits less variation of losses. In summer season and with maximum photovoltaic infeed, the disadvantage of the schemes a) and b) is obvious. Due to the restriction of providing only capacitive reactive power without considering the actual network voltage, voltage levels are still boosted even when the network voltage is already high. Therefore, the additional losses can increase up to 50 %.

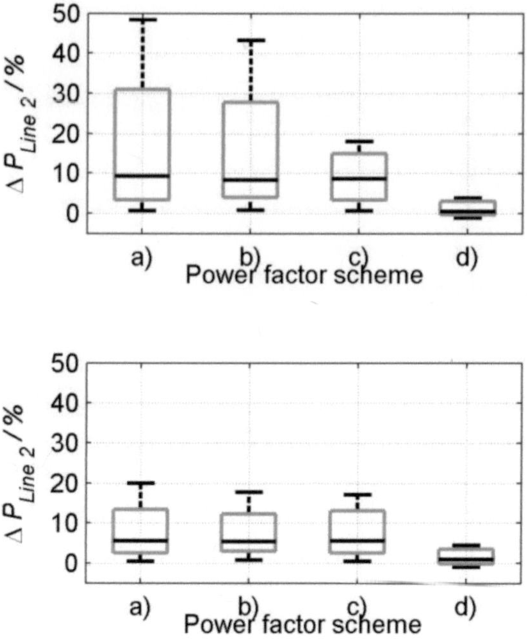

Figure 23: Additional losses of the four evaluated power factor schemes for the city scenario in summer (top) and winter (bottom).

Table 9: Medians of the village, suburb and city scenario in winter and summer season.

Scenario / Median	Village $\Delta P_{Line\ 1}$ / %	Suburb $\Delta P_{Line\ 4}$ / %	City $\Delta P_{Line\ 2}$ / %
Winter			
a)	10.1	3.1	5.6
b)	9.0	3.1	5.4
c)	10.1	3.1	5.6
d)	4.5	3.0	1.0
Summer			
a)	15.9	5.0	9.3
b)	14.6	5.0	8.3
c)	14.7	5.0	8.6
d)	3.5	3.0	0.5

As it can be learned from Table 9, the hysteresis controller (d) yields always the best performance, in summer as well in winter season, and in all three residential scenarios. The effect is especially advantageous for the Village and City scenarios and to a lesser extent for the Suburb scenario.

3.3 Conclusion

The requirements for grid access for passive rectifiers and rectifiers featuring an active front-end have been compared. DC-EVSE close to a local substation can be installed without special requirements. The benefits of active front-end become apparent as soon as line length increases. For distant DC-EVSE locations, the supply voltage limitation is crucial. Reactive power voltage support is an effective method to increase maximal line length up to 30 %.

A *grid-in-the*-loop concept has been developed in this investigation to assess the performance of different reactive power voltage support schemes, allowing DC-EVSE operation in remote network positions. Therefore, the algorithms for DC-EVSEs reactive power voltage support were adopted from DREs grid connection requirements.

Typical network topologies have been simulated on a load flow simulation, which had been coupled in real-time to a power converter system emulating DC-EVSE operation. The impacts on network losses and voltage stability have been evaluated. Considering scenarios with DRE infeed and scenarios without DRE, the hysteresis based reactive power voltage support scheme showed superior performance for voltage stabilization and can therefore be recommended for implementation on DC-EVSE.

3.4 References

Arruda, L., Silva, S. and Filho, B., "PLL structures for utility connected systems," in Industry Applications Conference, 2001. Thirty-Sixth IAS Annual Meeting, 2001.

Briz, F., Degner, M. and Lorenz, R., "Analysis and design of current regulators using complex vectors," in IEEE Transactions on Industry Applications , vol. 36, no. 3, 2000.

Bundesverband der Energie- und Wasserwirtschaft (BDEW, German Energy and Water Association), „TAB 2007 Technical conditions for connection to the low voltage network, Edition 2011," 2011.

CENELEC, Standard EN 50160:2010 + Cor.:2010 Voltage characteristics of electricity supplied by public distribution networks," 2010.

DIN German Institute for Standardization, „Standard DIN EN-61000-3-11:2001-04 Electromagnetic compatibility (EMC): Limits; Limitation of voltage changes, voltage fluctuations and flicker in public low-voltage supply systems; Equipment with rated current <= 75 A," 2001.

DIN German Institute for Standardization, „Standard DIN EN 61000-3-12:2012-06 Electromagnetic compatibility (EMC): Limits - Limits for harmonic currents produced by equipment connected to public low-voltage systems with input current > 16 A and <= 75 A per phase," 2012.

D-A-CH-CZ working group EMC, „D-A-CH-CZ Technical rules for the assessment of network disturbances, 2nd Edition," 2007.

E. C. Institute for Environment and Sustainability, "Photovoltaic Geographical Information System," [Online]. Available: http://re.jrc.ec.europa.eu/pvgis/apps4/pvest.php?lang=en&map=europe. [Accessed 2014].

Kerber, G., Capacity of Low Voltage Distribution Networks Due to Power generation of Small Photovoltaic Power Plants, München, Germany: Ludwig-Maximalilians-Universität München, 2011, pp. 22-26.

National Platform for Electric Mobility (NPE), "4th Progress Report, Joint Agency for Electric Mobility of the Federal Government (GGEMO)," Berlin, 2014.

Pflugradt, N., Teuscher, J., Platzer, B. and Schufft, W., "Analysing low-voltage grids using a behaviour based load profile generator," Renewable Energy and Power Quality Journal, 2013.

Scheffler, J., Bestimmung der maximal zulässigen Netzanschlussleistung photovoltaischer Energiewandlungsanlagen in Wohngebieten., Fakultät der Elektro- und Informationstechnik der Technischen Universität Chemnitz, 2002.

Statistical Office of the Federal State of Baden-Wuerttemberg, „Struktur- und Regionaldatenbanken," [Online]. Available: http://www.statistik.baden-wuerttemberg.de/. [Accessed 18 10 2014].

VDE Association for Electrical, Electronic & Information Technologies, VDE Application Rule AR-N 4105 Generators connected to the low-voltage distribution network., 2011.

Westnetz GmbH, Technical conditions for connection to the low voltage network, 2013.

Zimmerman, R., Murillo-Sánchez, C. and Thomas, R., "MATPOWER: Steady-State Operations, Planning, and Analysis Tools for Power Systems Research and Education," IEEE Transactions on Power Systems, pp. 12-19, 21 June 2010.

4 How professional and private individuals use and charge their EV

Magali Pierre, Anne-Sophie Fulda***

**EDF Research and Development, ICAME*
***European Institute for Energy Research*

This chapter presents results from 45 interviews in the CROME demonstration projects targeting employees on the one hand and private individuals on the other hand. The first part of the chapter insists on the professional challenges that determine the relationship to EV. The second part of the chapter is focused on users of EV that have purchased an EV as a personal decision and examines howtheir mobility practices was changing during the first year with this new mobility technology.

This chapter sets out the results of a sociological study focused on acceptance of EV conducted as part of the CROME project. "What explains the success of an innovation?" This is the fundamental question asked in 1988 by Madeleine Akrich, Michel Callon and Bruno Latour in a famous eponymous text (Akrich et al., 1988). It is indeed not only the quality or technical excellence of a product that will ensure its success and its dissemination, but rather the process by which this product has gradually engaged the support of some social groups and managed to "interest" the various actors: users and also engineers, designers, etc. We precisely described in a previous paper (Pierre et al., 2015) how a technical device contributes in engaging the participants in the experiment, constituting a spur to virtuous practices (Latour, 2000). In the present chapter we rather focus on the appropriation of EV and their integration into daily life routines. By analysing the appropriation of a technical system by users, we align ourselves with the field of sociology of usage and consumption, which "[examines] usages by relocating them in a general action" (Jouët, 2000, p. 499, *translated by us*).

Our study examines the mentioned challenges from the standpoint of how the vehicles and charging infrastructure are actually used by EV users. In this chapter, we focus on the results in France: how is the mobility of EV drivers reconfigured to cope with limited driving range? Whilst this is a standard question related to EV (s. Jarrigeon et al., 2014), we could not leave it aside in this experiment, especially as the diversity of testers (both professionals and private individuals) allowed us to analyse a wide variety of responses to this constraint.

4.1 Literature review

As explained in the literature review addressed by Rezvani et al. (2015, p.133), "it becomes (...) important to focus on the actual adoption behaviour concerning EV and not only on intentions" – contrary to the majority of the available studies (e.g. Lieven et al., 2011, Ozaki and Sevastyanova, 2011). Some researchers have explored to what extent the use of EV will modify its acceptation, and more specifically not the acceptability but the acceptance (Bühler et al., 2014, p.36): "Acceptance is one's attitudinal and behavioural reaction after exposure to a product. Prior to exposure, only acceptability can be assessed". Indeed, EV usage experience positively impacts EV purchase intentions because of high satisfaction of users (Ensslen et al., 2013).

Most of the inventoried literature on social aspects concerning EV concentrates on the purchase intentions (s. Ensslen et al., 2012 and 2015). However, we consider that in order to understand the reasons why an innovation succeeds it is much more promising to analyse the conditions of its dissemination and the current practices that surround this innovation than to adopt the point of view of purchase intentions.

Another point is that most of the studies generally deal with a few days use (Graham-Rowe et al., 2012), whereas taking into account long time period reduces obviously the bias effects that any trial test induces. Indeed "the issue of compatibility of EV in the everyday lives of consumers and their habits has also been found as an important contributing factor for potential adopters" (Rezvani et al., 2015, p.131). In other words upcoming research should involve long time periods, as "it seems promising to focus more on studies that examine acceptance of EV within the context of real-life experience" (Bühler et al., 2014, p. 35). This is also the case of the field studies that we will develop here. The main outcomes of few recent studies[8] that deal with a long term use of EV (several months) are depicted in the following.

This is precisely the subject of an article by Bühler et al. (2014) that shows the effect of the experience and even of an intensive use of EV on the acceptance. This survey analyses the perceptions of people who have experimented EV for six months in Berlin and indicates that this use promotes a very favourable opinion but doesn't determine purchase intentions, which depend on other factors. The authors classified the experiential benefits on one side and those who do not come from the use of EV on the other.

Furthermore we must mention the significant survey from Ryghaug and Toftaker (2014) about EV buyers. They applied a combined method based on practice theories and theories of domestication, which are interesting for understanding the long term setting up of the EV in a specific context (see also Pierre and Fulda, 2015). They focus on the incorporation of new practices in routines, given that appropriation is at the heart of anchoring an innovation among user groups. This appropriation of an EV is based on a learning process that creates a reflexive momentum for establishing new routines displacement.

Pierre et al. (2011) analysed the pioneers of EV in the early 2000s, those who could be called innovators in the theory of diffusion's terminology (Rogers, 1962). These people daily overcame the difficulties of the limited

---

8 E.g. the study (from Burgess *et al.*, 2013) lasted for a several month period. Nevertheless, the authors do not analyse their practice and usage of the vehicles (neither the declared ones) but the meaning that the EV carries for them and their opinion on it.

battery range through avoidance of certain journeys. Thus EV private users easily faced travelling difficulties and did not change their mobility patterns significantly. In the present chapter we intend to ask for the same questions among the present-day EV users, who could be identified as early adopters rather than innovators in the Rogers' terminology. Our study is one of the few to date that concerns private users who bought full EV, thus having made a significant investment in this kind of car[9]. They are therefore users not for a few weeks but for years. We are describing their coping with challenges that seem inherent in the use of an EV: how daily routines get installed, the constraints that this technology involves and how individuals overcome them, voluntary and conscious changes they had to make.

4.2 Data and method used

The study therefore concentrated on two specific populations who experienced EV and the charging infrastructure: employees of companies and public institutions on the one hand, and private EV owners on the other hand. As the project was gradually deployed in the Moselle region and in Germany, essentially among professionals (data collected in 2012), and then in the Bas-Rhin region, mainly among private owners (data collected in 2013), we interviewed the two sets of participants in two waves, as part of two successive field research projects with relatively clear cut-off points (cf. Table 10).

This research project consisted in a field study, with the objective of understanding how people used their vehicles and their opinion about the charging infrastructure. It involved face-to-face interviews of a semi-directive nature. During the interviews, our aim was to gain an overall vision of mobility and charging practices. More than twenty people were interviewed during each phase of the study.

[9] In this perspective, it is comparable to the studies from Ozaki and Sevastyanova (2011) and Caparello and Kurani (2012) about plug-in-hybrid vehicles.

Table 10: Description of the sample.

Field research 2012	Field research 2013
• Corporate and public institution Professionals and EV managers	• Private individuals and/or independent professionals
• Moselle (Lorraine)	• Bas-Rhin (Alsace)
• Data collected in 2012	• Data collected in 2013
• 22 interviews (30 people)	• 25 interviews
• No public charging infrastructure in place at the time of the study, but projects underway	• Public charging infrastructure being finalized at time of study
• Public incentive: national bonus+ distribution of a domestic socket + rebate on the leasing of the car	• Public incentive: national bonus on the purchase of the car + regional bonus conditioned to the check of an appropriate charging wall box (whatever type 3 or domestic socket)

These qualitative studies were part of the evaluation work package of the CROME project. They aimed at evaluating the acceptance of EV and charging process through a bottom-up method. This comprehensive and inductive method completed the online questionnaires carried out by the KIT (see chapter 8) by gathering the users' and fleet managers' spontaneous opinions on using an EV and charging practices: when, where, who, how often, what difficulties were encountered. It consisted of face-to-face interviews, presenting the users' interests, problems, needs and experiences. The interviews generally took place in the work place or at the users' home, which offered an opportunity to see the car, the electric installation, and sometimes to observe the users' behaviours with their EV.

This chapter is structured as follows: The first part of the chapter intends to analyse the various reasons why stakeholders (i.e. car fleet managers or end-users) take part in these trials. At the same time, it insists on the professional challenges that determine the user's relationship to his EV. The second part of the chapter deals with private EV users who have purchased a car as a personal decision and examines how their mobility practices were changing during the first year with the EV.

Although experimental conditions are generally designed to flatten the differences between individuals in order to highlight the relationship with a technical device "all things being equal", our analysis points out that user react differently. We therefore distinguish in our field study two user groups: the professionals (employees) using an EV (usually a fleet vehicle) and the private EV users, using their own car.

4.3 The use in a professional context

The first field study carried out in France (data collected in autumn 2012) dealt with professionals using EV when working for regional authorities or companies. We met occasional and regular users of EV, as well as fleet managers, who are responsible for EV within the aforementioned organisations.

During this study phase, we met 7 women and 23 men[10] in France (essentially in the Moselle region in mid-2012) in order to talk about the use or management of their EV. They had a wide range of jobs: some work for groupings of conurbations or municipalities (from secretary to general resources manager or technical service director) and some for companies in the energy sector (this sector was over-represented) and sometimes from the automotive sector. There were also some less typical professions, such as a director of a radio station and the manager of a passenger transport company.

4.3.1 The reasons for getting involved

For the organisations which employ these people, getting involved in the CROME project was a way of meeting their environmental obligations and of demonstrating their commitment to clients and citizens. The involved manufacturers– some in the automobile or electrical components sector – also stressed how useful it was for them to be able to test EV equipment or components, and how they intended to make savings by replacing part of

[10] Sometimes in individual interviews, sometimes in grouped interviews, with 22 interviews in total.

their fleets with EV (e.g. by lower maintenance and fuel costs). The local authorities placed greater emphasis on local development: maintaining employment (automotive component manufacturers in particular) especially within the automotive and electrical industry.

The purchase of EV by these organisations, sometimes in large quantities (up to 8 EV in a fleet), was partly initiated by the General Council, which subsidized an important part of the cost[11] and supported the project (promotion role and cooperation with project consortium). This support significantly contributed to a positive purchase decision from fleet test participants, during times of high uncertainty.

4.3.2 The evolutions of the opinion on the EV through usage

Users were very pleased to have a vehicle that has a good acceleration (at low speeds) and which is quiet and pleasant to drive. Such opinions often followed a period of reluctance, with training and practice being required for attitudes to change. The negative points which were mentioned, related mainly to the following items:

- the small size of the EV (which should be interpreted with respect to the mostly applied EV, i.e. the Smart ED and Peugeot Ion) and the lack of boot/storage space.
- weak acceleration particularly for overtaking and
- poor battery range and loss of capacity in cold weather (mainly due to heating of the driver cabin).

4.3.3 Allocation modalities in the companies

Generally speaking, within the companies, the EV were distributed on the basis of criteria of use, to persons or departments travelling short distanc-

[11] The support provided by the Moselle General Council consisted in financing half of the cost of the EV rental, organizing the distribution of charging stations in partnership with EDF, and promising to help with the purchase of a specific charging wall box.

es: directors of such organisations (helped by fleet managers) selected people and units supposed to travel short distances on a regular basis. This said, a detailed analysis shows that conditions of use varied, depending on the organisation concerned. For half of the face-to-face interviewee sample, the EV were part of corporate fleets shared by numerous employees; for the other half, they were used by one or two employees who coordinated their needs.

Shared use of EV (self-service electric vehicles (SEV)) by several users

Shared use of EV (SEV) is seen as a way of avoiding periods of non-use, i.e. idle-time where the required charging already ended. This mode of operation is more or less successful depending on the organisation and user access. For example, taking the EV home after work increases the appropriation of the car.

EV are sometimes used for managerial valorisation and are thus allocated in accordance with a complex plan which takes account of seniority or merit. For instance in large companies (with more than one hundred employees) that have a pyramid organisation and a strong concern for equity, the EV gets allocated in line with managerial criteria: the "worker of the week" might be allowed to take the EV home for the weekend.

SEV allocated to one or two person(s)

There is a certain amount of trial and error before companies find the best status for these vehicles, probably because there is no universal solution. Judging by the index for the number of kilometres travelled, it is among individual users that we find both the most promising (i.e. high number of kilometres travelled by people making numerous local journeys) and least promising cases (e.g. employees who work in one place and not out in the field). Therefore, the driving patterns should be analysed in order to assure a consistent allocation to convenient users.

A beneficial softening of the usage conditions

In both situations (SEV for single or several users) people might be allowed to take the EV home. Therefore, we analysed both situations within the CROME field test. These conditions increase the variability of our acceptance study. However, the possibility to take the car home increases the mileage of the EV[12] and its acceptance.

4.3.4 Recharging the EV, a user-friendly gesture

People appreciated to charge on-site at work. The procedure was seen as simple and allowed them to avoid the inconvenience of going to a petrol station. On-site domestic socket systems were the preferred charging technology for most users. It was distributed at the start of the field test. Vehicle charging was systematic in cases where the EV were shared (mainly after each trip), and generally more 'reasoned' when there was a smaller number of users – people waited until a certain threshold was reached before plugging-in the EV.

Although vehicles were mostly charged at work, charging sometimes took place on standard household sockets, particularly at employees' homes, when the home installation had been diagnosed as suitable. Home charging only took place occasionally, but could be more frequent when authorized by management in the case of employees who were familiar with electrics. Such flexibility was seen as a way to allow employees to familiarize themselves with this new technology and to allow occasional charging. Some people mentioned incidents which had occurred during home charging due to old electricity installations (e.g. extension cord burning during a long charging process).

There was little feedback on the public infrastructure, which was not sufficiently operational during this first project phase (in 2012). Neverthe-

12 This link was not totally systematic, as frequency of use and distance travelled must also be taken into account when explaining annual mileage.

less, we were able to see that employees rarely used public charging infrastructure in France and even less in Germany. In particular, people working for local authorities rarely need to travel to regions other than those where their citizens live. Nonetheless, some professionals asked for information about the allocation and technical specifications of future charging stations abroad and the possibility to charge their car– assuming a potential future need.

The previous part of our chapter dealt with the appropriation of EV in the workplace and underlined the importance of the spreading conditions of this innovation. The success in replacing combustion engine company vehicles by EV mainly depends on professional factors such as the flexibility of the usage conditions. In other words, the technical system does not impose alone its effects, but deployment facilities generate distinct practice modalities that are linked to diverse appropriations of the car. Professional deployment context has a structuring effect as it is shown by the possibility of coming home by this electric corporate vehicle.

In other words, EV users are not neutral average people, free of any experience and willing to candidly test an innovation whose technical characteristics would bear alone the reasons for its success or failure. Their use of the vehicle and of the charging infrastructure has to be interpreted in the context of the fleet test.

4.4 The use by private individuals

The second phase of the study focused on private EV users interviewed in autumn 2013 and dealt with how they used the EV and perceived the CROME charging infrastructure. The sample was selected from a dataset provided by Sodetrel[13] corresponding to EV purchasers who had asked for a public infrastructure access pass in the field test region.

[13] Sodetrel is an EDF's subsidiary entirely dedicated to charging infrastructure systems.

Approximately 25 interviews were carried out in 2013 among private EV owners living in Bas-Rhin. Their EV were used for private journeys, though in certain cases freelancers also used them for professional purposes.

4.4.1 Acquiring an EV: how and why

Among the private EV users who decided to make such a significant investment, we identified three purchase rationales. Some of them were developing a technology-oriented project which went far beyond this purchase. For instance a household lives in a smart-home and appreciates controlling all energy fluxes inside the house; these people see the EV as a kind of new electrical appliance that has to be managed. Others mentioned financial reasons, in particular the rise in petrol prices or the cost of repairing conventional vehicles. In all cases, and for the latter in particular, the regional subsidy was the deciding factor, turning an interesting opportunity into a bargain not to be missed[14]. Ecological factors tended to remain in the background; some people mentioned the desire to restrict noise in towns or to reduce CO_2 emissions as leading to increased interest in EV, but for most people whilst these were certainly underlying arguments, they were not decisive factors when deciding to purchase an EV.

The choice of which model of EV to buy was generally based on classical criteria: the presumed quality of the vehicle in terms of size, vehicle range, comfort, aesthetics, etc. However the question of charging standards was one of the criteria at the time of purchase: for some buyers the fact of having a dedicated charging station put them off buying certain models and led them to choose an EV which could be charged on a standard socket. Another important criterion was the type of acquisition (leasing *versus* purchase). Whether they preferred to rent the EV (and more specifically its battery) or buy it, this aspect was highly important for some people. It

[14] In Pierre *et al.*, 2011, the buying motives referring to pioneering-ecological spirit (combination between current ecology and technophilia) and to seizing opportunities (generous subsidies, bargains coming from relatives) were highlighted.

might be interesting for future research to link this concern to a general conception of investment in sustainable goods.

4.4.2 Installing a charging wall-box at home

The wide variety of home charging models (e.g. by Hager, Schneider and Sobemscam) and types (type 3 socket, domestic socket, etc.) proves that the CROME experiment took place under real market conditions. The cost of the private charging infrastructure and the installation procedure are two points that were generally not considered in advance, and which were thankfully cushioned by the perspective of a conditional subsidy: those who bought a dedicated Type 3 socket had not foreseen the cost, and it would appear that the conditional subsidies provided by local authorities played a large part in this price being accepted. Some people felt that the private charging points should be acquired for free, or included in the price of the EV, except in cases when such an installation would be a chance to improve the service by accelerated charging for example, consumption readings or guaranteed electrical safety. When selecting a charging point supplier and installer, the criteria were the installation cost, reputation in terms of technical quality and local preference. The high installation cost caused some customers to choose local electricians (small companies) who were cheaper and who could advise them on choosing a home station supplier.

With respect to home charging infrastructure, the interview sample helped to identify a few tricky cases: people who rented their garage, or who had no private parking space, or whose garage had co-ownership status. Such cases were not as problematic as they might seem. Generally speaking, for people renting a house with garage, the installation of a charging point usually does not pose a significant problem in France if the owner of the garage agrees and the EV user is paying for the installation.

The issue of people living in an apartment with a shared garage is more complex, particularly because the costs for electricity consumption in the garage or at the outside-sockets is usually equally distributed among all apartments. For co-owners, the difficulties are threefold: first of all they need to persuade the property management company or co-ownership trustee of their right to install a charging wall-box. Second the installation of the wall-box including metering is more complex. Third, once the wall box is installed, the apartment is often far from the garage; so either it is possible for people to be connected to their own meter though it is not always possible to lay the cables. If this is not possible, they must negotiate with the joint tenants to pay their specific share of the electricity used. Customers whose garage is an integral part of the building will need professional support, legal, technical and communicational aid from their energy company for these various matters. Some people did indeed say that they felt helpless with regard to the technical choices they had to make.

Finally, those who did not have a private parking space used public spaces to charge their vehicles. There were 3 such people in our sample. We will explain how they used the public charging infrastructure further down (in part 4.4.5).

4.4.3 Charging the EV: a reflexive routine

Home charging was done on a routine basis without any problem. Most private individuals integrated charging into their daily schedules, mainly charging at home and often in the evening. Most of the users we met practiced what we might call *reasoned charging*, i.e. they charged their vehicles when they felt it was needed and not systematically as soon as the EV was at a charging point. Hence, the mileage defined the charging frequency, which generally varied between 2 and 4 days, depending on the user. This charging frequency was calculated on the basis of the remaining battery range in relation to the journeys to be made. In other word the charging routine relies on quite a complex assessment that is embedded in each EV user's routine gestures.

It is worth noting that the majority of users were aware of the charging cost, which they deemed to be relatively low; they were very confident about the low levels of electricity consumption. They were interested in having specific consumption details. There were numerous questions about charging, particularly concerning the possibility of using an extension cable, the EV's energy consumption and the modes of accelerated and/or fast charging.

4.4.4 The EV, an everyday car

Because EV only have a limited driving range, due to the battery capacity, it is fair to ask whether users feel free of their movement and not restrained by the EV capacities. An analysis of the declared journeys – and those which could not be made – provides us with information on this matter: EV very often became the family's primary car, and in some cases their only car. According to our interviews, some people travelled relatively large distances. We agree here with Revzani asserting that "limited driving range is more of a perceived barrier than an actual one" (Rezvani et al., 2015, p.130).

Most of the users we met use their EV for regular journeys over short or medium distances. Only some journeys by EV were relatively long-distance trips (around 100 km). People make surprisingly precise forecasts of their future trips. In the users' mind an EV is well-suited to regular medium-range journeys, because it is an everyday vehicle, particularly for people used to rural and suburban regions. The EV is the most used vehicle, with family members competing for car usage. Of course, it is not yet easy to use an EV for long-distance or unplanned journeys. The interviewees tended to use another vehicle for such journeys.

Our socioeconomic results from the CROME field test are in line with the main outcomes about changes in driving behaviour by other researchers. For instance, some researchers working on the plug-in-hybrid EV (Ozaki et al., 2013) analyse "how drivers' existent knowledge and competencies are mobilized and transformed while developing the skills required to use a hybrid car" (Ozaki et al. 2013). We confirmed their results in many aspects of environmentally improved driving behaviour: people slow down their speed, anticipate stops earlier, watch consumption per 100 km on the dashboard, accelerate less frequently and increase the time of constant speed.

To cope with unexpected situations with their EV, users adopt charging 'emergency' solutions: they use fast charging at public stations or household sockets in the social neighbourhood. This type of recharging remains infrequent, but is very much appreciated by users, as it increases their driving range. In addition most users, and not just those who do not have garages, appreciate that the public space provides a charging service open to all electric cars. From this point of view the need for public charging points is undeniable.

4.4.5 The use of the public charging infrastructure

A typology of the users

Our sample can divided in terms of intensity of use of the public accessible charging points on-street, in car parks and in supermarkets:

First, a small share of interviewees does not use these public accessible charging points and are unaware of how to use them. They live far from Strasbourg and never go there, or else they live in the city but have a garage and so do not need to charge elsewhere.

Second, another small share of interviewees use these charging points almost every day or at least several times a week; they use stations located either at work or near where they live in which case their home is in a building, in the town center, without garage. In such cases it is a real necessity: 3 families in our sample had no private charging station and therefore could only charge their EV in public spaces. The ability to charge their vehicles this way was a decisive factor when deciding whether or not to buy an EV. They almost always use the same charging point, as it perfectly suits their destination. Having a small number of preferred points suffices to exclusively charge at public stations.

Finally, for the majority of users, who charge occasionally, public charging is a useful option. One of the reasons for using public stations is the charging speed: occasional users like to have access to an accelerated charging point, and consider this to be a major advantage of public points. They appreciate to benefit from the fast charging at a spot where they go for their work, shopping or visits to the town center. This is the same motive as those given by interviewees who use public points more frequently. They like the accelerated charging which is used according to individual needs and the ability to "make the most of" exceptional charging conditions.

Expectations on the public charging infrastructure

Negative feedback concerning the public charging infrastructure focused on:

- malfunctions: extension lead like stuck in a socket,
- question of standards: terminal not always compatible with the EV model,
- access times and conditions: particularly at supermarkets
- need for information for instance oncharging point availability.

Regarding the problem of compatibility and the need to be able to use all charging points, most users were aware that they are in a very young marketbut all users indicated their wish, that in the future all charging stations should accessible for each EV user. Regarding the other issues, users are waiting for the various involved institutions and companies to propose solutions and explanations, notably via a hotline which would be permanently available. The majority of users are waiting for the following improvements:

- updated information on the location of public charging points,
- information about the compatibility of charging points,
- 24-hour access and
- increased maintenance quality.

The expansion of the public charging point network is seen as a condition of use, as is reliability – and even impeccability.

The cross-border issue was important for the people running the CROME project, but not so much for the users, who dealt with real situations by using a conventional car, or else an EV which does not need to be charged abroad as it was possible to do the return trip without charging.

Regarding accelerated or fast charging in public zones, there are specific expectations relating to the turnover of EV which are charging so that the charging locations are always available. It should also be noted that some people are confused about the difference between normal, accelerated and fast charging.

Finally, users wish to see improvements to the devices used to locate charging points, and their remarks provide an outline for the ideal device: it should be multi-channelled: on the Internet, on a non-proprietary smartphone application, and on a paper list, be regularly updated, of the types of socket and vehicle compatibility, and the availability of the charging points in real time. Furthermore, improvements are required with regard to the information EV users receive at the time of purchase. The EDF application (http://m.sodetrel.fr/crome/) meets all these needs but few people are aware of it, whereas the Chargemap collaborative website (http://fr.chargemap.com/) is well-known but is not always reliable.

4.5 Discussion: main challenges from the users' perspective

We now discuss the main issues of inter-operability and public charging acceptance of the CROME project, and summarize the answers that our field study has provided.

4.5.1 Battery range in areas with high automobility

Professionals seem to easily cope with the issue of battery driving range. Whether for purely professional trips or for commuting, vehicles are allocated in terms of the battery range, to avoid running into any borderline situations. As far as they are concerned, commuting is one of the EV's main usage patterns.

Even though some private individuals are unhappy with the limited battery range of their EV, cross-border trips are rare. Some families still have a conventional car which allows them to go on longer journeys, whilst those who no longer have a conventional vehicle are always able to get hold of one by borrowing or renting. They nevertheless still want reliable charging from public stations, this being seen as a condition for extending their range of EV journeys.

4.5.2 Diversity of technologies and different standards in France and Germany at the start of the project

Did the different technologies tested in CROME work? Who/what is blamed for the dysfunctions and what do people think about the way these are resolved? How do users perceive the issues of compatibility between the different EV technologies and the charging infrastructure (Gagnol et al., 2013), charging interoperability and standardisation?

With the exception of a few specific EV breakdowns and certain critical cases of charging in public places, the professionals we interviewed deemed the technical devices tested (vehicles and private charging infrastructure) to be satisfactory. The proper operation of EV and their proper coupling with the charging infrastructure were considered to be implicit and were not underlined by users. Issues of compatibility and standards were considered to be absolutely crucial by private users who had to cover larger areas than the professionals. The question of interoperability from one country to another, including the cross-border network, was also important for private individuals, as the majority does not exclude the possibility of one day using the infrastructure available at a European scale.

The questions of standards for private charging stations were an issue for private users, many of whom criticized the fact that partnerships between manufacturers has led to captive technologies. Additionally, they do not understand why certain EV require a specific charging station (type of socket), or why home wall-boxes are so expensive.

4.5.3 Cross-border travel

Are there users who regularly travel back and forth between countries? And are they worried about being able to charge their EV on both sides of the border? Are behaviours/opinions the same in France and Germany? What is the extent of the infrastructure in the different zones?

Whilst professionals rarely travel to Germany for business purposes, some of them would like to do so (for private reasons). Furthermore, knowing this to be possible would be an additional reason for buying an EV for personal use. Private individuals expressed an even clearer desire in this respect, but also deeper concerns. Some were especially suspicious with regard to charging abroad, given that fewer alternatives are available to them in foreign countries.

4.6 Conclusion

In what ways do the uses of EV and charging infrastructures differ between professionals and private individuals? As we have just seen, the results provide various elements of responses relating to the differences between professionals and private users, whether regarding the perception of EV, charging methods, financial investments or the feeling of participating in projects for the future.

In both cases (professional a private EV motorists) we showed how relevant usage conditions and concrete practices can be to understand how an innovation gets appropriated by end-users. For the latter, we underlined the importance of access conditions to the private and public charging infrastructure. The part that dealt with EV in a professional context drew usage patterns (the so-called "cadres d'usage" from (Flichy, 2007)) that lie at the interface between two kinds of prescriptions: the technical script ordering to use the EV mainly for short journeys and professional requirements that mediate the organisational conditions of using an EV. Exploring this socio-organisational system helped us to go beyond the traditional positioning of the sociology of usage centered on the interaction between a user and a technical device.

4.7 Acknowledgements

We would like to thank Patrick Gagnol from EDF for its lasting and stimulating support to this study. Furthermore we thank the research team of KIT (especially Axel Ensslen, Mathias Pfriem, Felix Vogel and Patrick Jochem) for interesting and productive exchanges during the project.

4.8 References

Akrich M., Callon M. et Latour B., "A quoi tient le succès des innovations ?", in Gérer et comprendre, Annales des Mines, 11, 1988, p. 4-29.

Alter N., L'innovation ordinaire, Paris, PUF, 2000.

Burgess M., King N., Harris M. and Lewis E., "Electric vehicle drivers' reported interactions with the public: Driving stereotype change?", Transportation Research Part F 17, 2013, p.33-44 .

Callon M., "L'Etat face à l'innovation technique. Le cas du véhicule électrique, Revue française de science politique", 29ème année, n°3, 1979, p.426-447.

Caperello, N.D. & Kurani, K.S., "Households' Stories of Their Encounters with a Plug-In Hybrid Electric Vehicle", Environment and Behaviour 44, 2012, p.493–508.

Ensslen A., Paetz AG., Babrowski S., Jochem P. and Fichtner W., "On the road to an electric mobility mass market – How can early adopters be characterised?", Workshop Future Mobility, Oldenburg, 2013.

Ensslen, A., Babrowski, S., Jochem, P., Fichtner, W., "Existe-il des différences d'acceptation des véhicules électriques entre la France et l'Allemagne? – Premiers résultats de l'analyse scientifique du test de flotte Cross Border Mobility for Electric Vehicles (CROME)". Proceedings of the 11ème séminaire francophone est-ouest de socio-économie des transports, Karlsruhe, 2012.

Ensslen, A., Jochem, P, Rometsch, M., Fichtner, W., "Adoption of Electric Vehicles in the French-German Context". In: Schäuble, J., Jochem, P., Fichtner, W. "CROME", 2015.

Flichy, P. (2007), Understanding technological innovation: A socio-technical approach, Edward Elgar, Cheltenham, UK.

Franke T. and Krems JF., "Understanding charging behaviour of electric vehicle users", Transportation Research Part F, 21, 2013, p.75-89.

Gagnol, P., Jochem, P., Pierre, M. & Fichtner W., "CROME: the French and German field Demonstration of the interoperable Electro-mobility with EV", Proceedings of EVS27-Conference, Barcelona, 2013.

Gjoen H. and Hard M. "Cultural Politics in Action: Developing User Scripts in Relation to the Electric Vehicle", Science, Technology and Human Values 27 (2), 2002, p.262-281.

Graham-Rowe, E., Gardner, B., Abraham, C., Skippon, S., Dittmar, H., Hutchins, R., & Stannard, J., "Mainstream consumers driving plug-in battery-electric and plug-in hybrid electric cars: A qualitative analysis of responses and evaluations", Transportation Research Part A: Policy and Practice 46, 2012, p.140–153.

Heffner R., Kurani K. S. et Turrentine T., "Symbolism in Early Markets for Hybrid Electric Vehicles", Institute of Transportation Studies, UC Davis, 2010, can be accessed via internet.

Jouët J., "Retour critique sur la sociologie des usages", in Réseaux, vol.18, n°100, 2000, p. 487-521.

Latour B., "La fin des moyens", Réseaux, volume 18, n°100, 2000, p.39-58.

Lieven, T., Mühlmeier, S., Henkel, S., & Waller, J. F., "Who will buy electric cars? An empirical study in Germany", Transportation Research Part D, 16(3), 2011, p.236-243.

Ozaki R., Shaw I. and Dogson M., "The Coproduction of 'Sustainability': Negociated Practices and the Prius", Science, Technology and Human Values, Vol.38 (4), 2013.

Ozaki, R., & Sevastyanova, K., "Going hybrid: An analysis of consumer purchase motivations", Energy Policy 39, 2011, p.2217–2227.

Pierre M. and Fulda AS., 2015, "Driving an EV: a new practice? How electric vehicle private users overcome limited battery range through their mobility practice", ECEEE Summer Study Proceedings, pp.907-916.

Pierre, M., "Script technique et prescriptions organisationnelles. Utiliser un véhicule hybride rechargeable en milieu professionnel", in Cihuelo J., Jobert A. and Grandclément C., Energie et transformations sociales. Enquêtes sur les interfaces énergétiques, éditions Lavoisier, Paris, 2015, p.51-67.

Pierre, M., Jemelin, C. & Louvet, N., "Driving an electric vehicle. A sociological analysis of pioneer users", Energy Efficiency journal, vol.4, 2011, pp.511-522.

Rezvani Z., Jansson J. and Bodin J., "Advances in consumer electric vehicle adoption research: a review and research agenda", Transportation Research Part D., 34, 2015, p.122-136.

Rogers E. M., Diffusion of Innovations. New York: Free Press of Glencoe, 1962.

Ryghaug M. and Toftaker M., "A Transformative Practice? Meaning, Competence, and Material Aspects of Driving Electric Cars in Norway", in Nature+Culture vol.9, number 2, 2014.

5 Further services for EV in the cross border context

Felix Vogel, Hartmut Schmeck

Institute of Applied Informatics and Formal Description Methods, Karlsruhe Institute of Technology

Electromobility services are important and necessary for increasing the attractiveness and pushing the development of electromobility on an international level. Numerous services are available in both, France and Germany, and they already support users in getting information about urban infrastructure and the management of electromobility. Some energy providers distribute authentication cards to users allowing them to access EVSE on both sides of the border. The interoperability of charging and payment services between both countries has been accomplished based on a roaming layer developed by Bosch. Fleet and private users from both countries can call e-services via Internet or mobile phones before, or access them via navigation systems during driving. The usability of the e-services has been studied using an online questionnaire. Overall, the results show that the users are satisfied with most of the services, but that they have not used them frequently. Development of electromobility services

Mostly, users need electromobility-services for the organisation and management of charging the batteries of their EV. Services in this scope have been developed by many providers for different functions and made available via Internet or mobile communication. In the CROME project the concept of services is focused on interoperability of systems between France and Germany (ISST, 2010). In the next subsections some aspects of development and realization of services by the KIT and the partners are presented.

5.1 Services in the CROME project

Within the CROME project various services have been designed and developed, particularly for charging and mobility with EV. KIT and the industry partners together came up with a set of required services and divided them into two categories: basic and additional services. In the first part of the project, the focus was on developing and testing basic services in both countries and on making them available to end users.

5.1.1 Basic and additional services

The basic idea of most services has been developed and differentiated during meetings between industry partners and the KIT. Many aspects have been discussed and the participants decided which services should be supported in the project. The basic services have the purpose to guarantee mobility and charging availability on both sides of the border, independent of different energy providers and EVSE.

The following basic services have been developed:

- searching for EVSE,
- simple (interoperable) charging and controlled charging,
- emergency solutions (e.g., in case of lost RFID card),
- personal usage information,
- delivery and cancellation of identification card,
- reporting of spot utilisation and customer behaviour.

A major part of these services could be implemented by software, a small part additionally required the customization of hardware. The additional services are intended to improve and support electromobility outside or inside of the cities.

The few value added services are listed below:

- preconditioning of the EV during parking,
- advanced controlled charging (timing and load),
- reservation of charging stations,
- direct payment,
- fleet management.

Both categories of services are available for fleet and private users participating in the CROME project. The majority of the properties of services were implemented by different industry partners in close cooperation with the KIT. The data communication between the backend servers and different services is encrypted and realized using a roaming layer.

5.1.2 Services for charging at a public EVSE

Many public EVSE are situated within cities (e.g., near the center or in shopping arcades) or close to cities (e.g., in the suburbs). In crome, a maximum of two electric cars can simultaneously charge at a public EVSE. The users need an RFID card for authentication, distributed by the respective provider. The card is also used for payment of energy consumption and/or for parking fees, depending on the contract between the partners (Standardization, 2011).

A display integrated in the EVSE provides several types of information: the price for different charging sockets per hour, the time for the charging process, provider information such as green vs. conventional energy, etc. (see Figure 24). The time from the beginning to the end of each charging phase has been recorded and sent to the backend server of the provider. The data is used for the billing process and for other analyses. With the opportunity of interoperability of services, each energy provider was able to implement different business models (Stadtraum, 2001). Some energy providers, particularly EnBW[15] in Germany and EDF[16] in France, developed additional service models and processes and offered them on their websites like: apps for finding public EVSE, or charging electronaut card.

Figure 24: A public EVSE with two recharger cables (left), and the display enlarged with socket information (right).

Figure 25: On the right side of the dashboard, the value of different services such as SOC, range capability, battery capacity, range in the city and charging time are displayed.

---

[15] Information of services from EnBW via internet. Source: https://www.enbw.com/privatkunden/tarife-und-produkte/e-mobilitaet/enbw-elektronauten-prepaid-ladekarte/index-3.html (accessed on 2014/10/14)

[16] Different Services from EDF via Internet. Source: http://about-us.edf.com/strategy-and-sustainable-development/our-positions/electric-mobility/experiments/electric-vehicle-282898.html (accessed on 2014/10/14)

5.1.3 Services inside an EV

Car manufacturers such as BMW, Daimler, Mitsubishi, Porsche, Renault, Toyota or VW are developing EV for private and business users. The cars have different electric properties like driving range, battery capacity, energy monitoring, performance, and charging speed. Inside EV, there are many sensors to catch all the activities of battery energy and states of the car during driving or the charging phase. These parameters are processed and monitored in a user-friendly way on the vehicles' dashboards. Among the most important parameters is the *state of charge (SOC)* value, it indicates the current charging state of the battery (in percentage points) (Wei He, 2013). Estimating the state of the battery is complicated and it depends on different factors such as battery construction, weather, car speed and the according algorithms. Nevertheless, this feature is essential for practical use of EV. Figure 25 shows an example of how information about the SOC and related data can be displayed on the dashboard of a car.

The drivers of EV appreciate information about driving range and charging time. The first service depends on parameters such as user road handling, traffic, weather, and battery capacity in a complex way. Most car manufacturers participating in the CROME project offer these services, and support planning the routes to the next charging station. Some services are available for mobile devices and can be accessed on both sides of the border.

5.2 Development of services

In the CROME project, several services have been designed and implemented. The purpose of the implementation was to develop alternative solutions for finding and reserving EVSE on both sides of the border. A simulation to calculate the shortest and fastest routes between occupied charging stations has additionally been implemented. The next subsections introduce the implemented application which can be accessed via Internet.

5.2.1 Application for locating EVSE

Most EVSE are placed inside or close to cities by energy providers or car park owners. The address of each charging station is stored in a database by the energy providers (e.g. EnBW and EDF). Both providers developed several web services only for contract partners. With the introduction of a roaming layer platform implemented by Bosch[17] the database was made accessible for the CROME project partners.

A web application can simplify the search for EVSE via Internet. Such a web application[18] has been developed by KIT. It displays the addresses and positions of requested charging stations. The search term can be entered into an input box or by clicking on the marker map. Using the input box, the user writes a city name which opens a list with the related addresses of EVSE. Selecting an address initiates the retrieval of many properties of the respective EVSE. These are displayed in a new frame on the right hand side on the window (see Figure 26).

The EVSE can inform the user about properties like socket type, price per time, provider name, country, and status – free vs. occupied. Below the frame, a street map is opened showing two markers; one for the current position of the user (red) and the other for the position of the free EVSE (green). Another function is to retrieve information about all EVSE within a city by clicking on one of the green markers on the left map.

The usability of the service is simple and it has an intuitive design, compared to other sites (ADAC, 2014). All functions are displayed within one window and can be read without scrolling. The application has been developed by the KIT and used modern web technologies, like PHP 5, HTML5, JQuery and CSS 3.

---

[17] Bosch eMobility Services. Source: https://bosch.bosch-emobility.com/de/com/home/homepage.html (accessed on 2014/10/19)

[18] KIT web application for finding electric vehicle supply equipments. Source: https://felixv.de/crome/html/index.php (accessed on 2014/10/19)

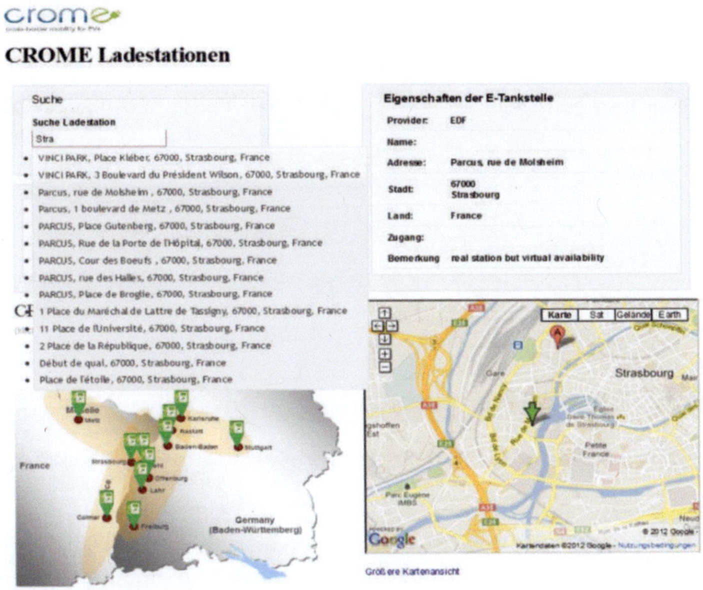

Figure 26: Web application for finding the address and position of EVSE, based on data from the roaming layer. (Source: https://felixv.de/crome/html/index.php)

5.2.2 Reservation of EVSE via web app

The reservation of public EVSE has been considered as an additional service in the CROME project. The number of EV in the cities increases every year, but the number of public charging stations remains moderate. Every energy provider can check if a public EVSE is busy or free and, accordingly, inform the user. For this purpose a reservation application for mobile devices has been developed by the KIT.

The development started with the analysis of requirements and design options. In cooperation with Bosch, groups of possible target groups of users have been identified as well as three kinds of reservation options (parking place, charging spots or sockets), and their location properties (public, semi-public or private). A few reservation scenarios for fleet and private users have been checked, considering all parameters such as parking time, parking spot availability, energy demand, and charging time (Lee, 2011).

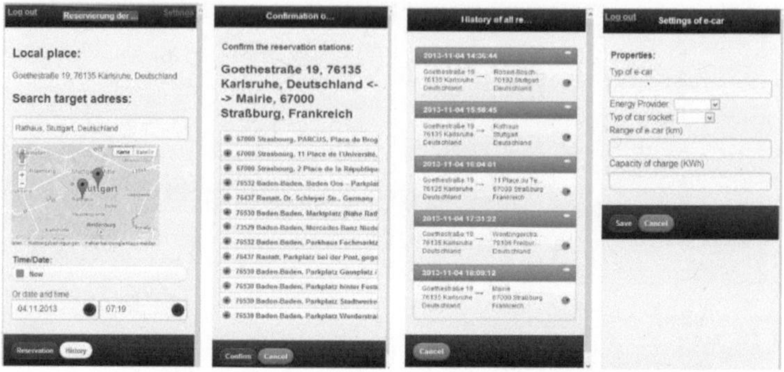

Figure 27: Web-application for reservation of EVSE in the CROME project.

Many use case scenarios have been introduced in the application shown above (see Figure 27). The application has been implemented for mobile devices and can be accessed via Internet. It is a light-weight application and can be used for the reservation of public charging stations in both, France and Germany. The user can enter the address of an EVSE into the search box and the preferred arrival and departure time and date. After the confirmation the system checks if the EVSE is free. If positive, the user can confirm the reservation of the EVSE which will be stored in the backend. Additional functions in the application are storing the histories of addresses visited before and filtering properties used in the past. The usability of the web-application is simple and intuitive.

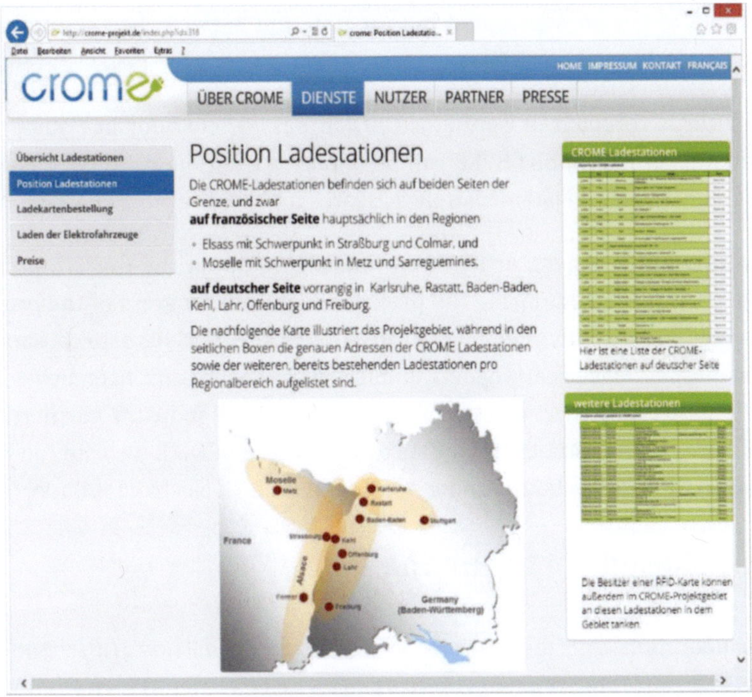

Figure 28: CROME-EVSE overview on ourwebsite, German version (www.crome-project.eu).

5.2.3 Services on CROME website

At the beginning of the CROME project, relevant information has been collected, processed and displayed on the CROME website. It started with the vision of the project and general information about work packages of the industry partners and the KIT. With European standardization of public charging stations like "Dual Type Socket" (Type-2 in Germany and Type-3 in France) it is possible to charge on both sides of the border with a personal charging cable. The authentication at a public EVSE on both sides of the Rhine is simple and it can be performed using an RFID card received from the respective energy provider. All charging stations are presented on the CROME website including their geo coordinates and their technical specifications (cf. Figure 28). In context of "International Roaming", it is possible to get an aggregated bill for different services like energy cost, basic or additional services.

RFID cards (prepaid in Germany, CROME pass or corridor in France) for authentication at public EVSE can be obtained from the respective energy provider. The card is immediately activated and available for charging.

Many research papers, articles and interviews about the CROME project have been published during the project phase. The progress of the project was regularly documented on the CROME website. Various aspects such as user mobility, fast charging, economical instruments and user behaviour have been analysed und tested by the KIT and the industry partners. All this information and the services provided by the CROME website can help the CROME users to better understand all aspects of electromobility.

5.3 Results of the online survey

In cooperation with the *Institute for Industrial Production (IIP)*[19] and the *European Institute for Energy Research (EIFER)*[20] at the KIT, two bilingual online questionnaires have been designed. The main purpose of the survey has been an analysis of the users' mobility behaviour with EV and the usability of e-services in France and in Germany. Both questionnaires have been provided in French and German (using a preferably equivalent phrasing), and addressed to both fleet and private users. The results of the analysis regarding the CROME services are discussed in the next sections.

5.3.1 Central questions of the study

In the project, the EV are used with different intentions and requirements. The users' behaviour with respect to mobility and the usage of services were central questions of the study. The survey contains questions about management, economy, technology, and infrastructure from a user's per-

[19] IIP website: http://www.iip.kit.edu/english/index.php
[20] EIFER website: http://www.eifer.kit.edu/

spective. Additionally, the experiences and expectations of users regarding EV and EVSE have been collected and sorted.

5.3.2 Construction of the online questionnaires

At the beginning of the questionnaire, there is an introductory text including a short description of the project, the intention and the structure of the questionnaire as well as further instructions and the expected duration of completion. For CROME participants, there are two text fields (user ID and EV ID) for identification. The first questions check if the participant owns an EV and the number of cars within his or her household (see Figures 29 and 30).

The remaining questions focus on the properties of the users' EV, reasons for purchasing an EV, performance of using an EV in every-day life and for work, and users' experiences with the charging process at (public) EVSE. Furthermore, questions about the usability of the RFID card, roaming, the search services (including searching for public EVSE, parking space fees, energy provider information etc.), and willingness to pay for electromobility are asked (Ensslen, 2013 and section 7).

Figure 29: Part of the online questionary in German.

The survey ends with two categories of questions about electromobity behaviour of the users and sociodemographic aspects. Both categories are important for analysing the data based on the environment of the users and the usability of electromobility. The last page provides text fields for general feedback regarding the survey or overall improvements, as well as further information about the KIT.

Figure 30: Part of the online questionnaire in French.

5.3.3 Usability of services by fleet users

The energy providers have already built a quite high number of public EVSE in the cities. These, together with several e-mobility-services facilitated the use of EV. The services are developed for different devices and they help users to find the closest public or semi-public EVSE during a trip.

Fleet users belong to companies and share one or more EV with other fleet users during a day or a week. They use the EV for business trips, e.g., for driving to customer sites or to transport documents or other things between different places. As the EV have to be continuously prepared for a trip, 60 % of the EV belonging to fleet users were daily charged on the premises.

To enhance electromobity, particularly fleet users desire fast EVSE at many different places. The places in France and Germany are shopping malls, motorways, main roads, city car parks and premises. Less frequent places are private homes or park-and-ride lots (see Figure 31). The benefits for the users include, of course, less worrying about charging and time saving (Ensslen et al., 2013).

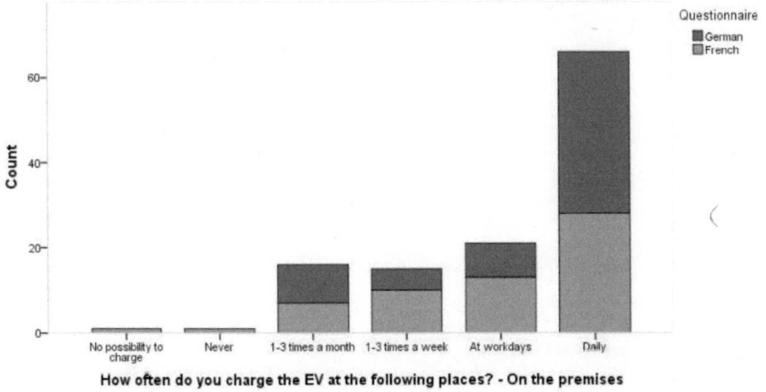

Figure 31: Chart depicting preferences regarding charging on the premises in Germany and France.

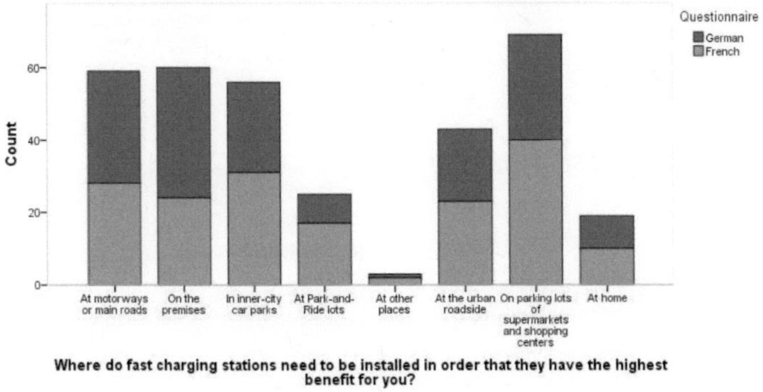

Figure 32: Chart depicting the preferences of desired places for fast charging stations.

5.3.4 Comparison between private and fleet users

In the final phase of the CROME project, the focus has been on analysing the management of services for private and fleet users in France and Germany. Both user groups have different electromobity behaviour and service requirements. Private users are more flexible than fleet users regarding the take-off-time of an EV. Fleet users, on the other hand, have strict time schedules and perform business trips of variable lengths.

The availability of fast EVSE at many different places is very important to both user groups, and a guarantee is desired that the charging and parking spots are free at arrival time. The reservation of public EVSE is an instrument for planning a trip for a long distance (see Figure 33). To the German users it is important to have a good development of electromobility infrastructure along main streets. A web-based system displaying all public charging stations is relevant to both groups, but not mandatory.

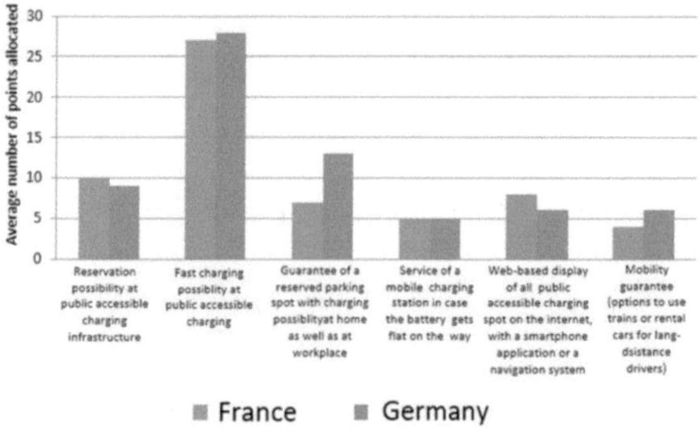

Figure 33: Chart depicting the importance of different services for private users.

The preference of German fleet users for fast charging at public stations in cities and the outside of cities is 15 % higher than the preference of French users (see Figure 34). French users prefer finding the closest charging station via internet and smartphone applications. A guarantee of a free parking spot at home and at the workplace is for German users very important. For long business trips, French EV drivers appreciate the reservation of public charging stations more than German EV drivers.

Both groups set their priorities rather diversely and selected a few e-services for their own preference. In France, the majority of people live outside the cities and EV are used more frequently than in Germany, particularly to approach the city center. A guarantee for free EVSE at any time is important to both groups, with the details depending on the exact situation and environment of the user. Private users are willing to use services for finding EVSE, as opposed to fleet users. In France, the users are more enthusiastic about using web-based technology for accessing services than in Germany (cf. Sammer, 2008).

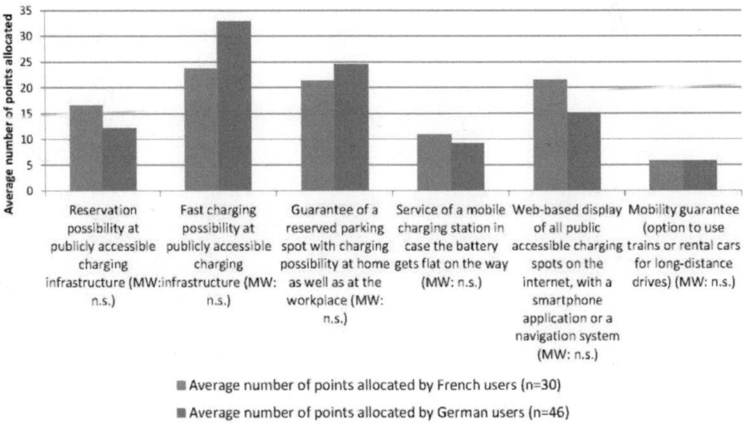

Figure 34: Chart depicting the importance of services for fleet users.

5.4 Conclusion

At the beginning of the CROME project, a large number of EVSE has been installed at various places in the cities and surrounding areas in France as well as in Germany. With an RFID card available from the respective energy provider, a user participating in the CROME project is able to authenticate at an EVSE and initiate charging. Various information such as the price of charging, the expected charging time, the availability EVSE, provider data, etc. can be retrieved directly at the EVSE. Depending on the parking infrastructure, the price for charging and parking can be summed up and directly paid via RFID card. On the website of the corresponding energy provider, the user can get a history of all charging activities and statistics concerning his or her usability of the e-mobility-services.

The interoperability of services on both sides of the border has been managed using a roaming layer platform developed by Bosch and other industry partners. In cooperation with Bosch, two web applications for finding and EVSE have been developed at KIT. The user interface appears to be friendly and intuitive. The application can be accessed via internet or mobile communications using various devices.

Fleet and private users evaluate the electromobity services quite well, regularly accessing them from different devices. Remarkably, French participants usually use the services during driving, accessing them via the navigation system of their car, while German users organize their trips in detail, using websites or mobile applications before going on a trip.

5.5 References

ADAC. (2014). SmartTanken. Source: http://www.smarttanken.de/ (accessed on 20/02/2014)

Ensslen, A. Jochem, P. and Fichtner, W. (2013). Experiences of EV Users in the French-German Context. Proceedings of EVS27-Conference. Barcelona, Spain.

Ensslen, A., Jochem, P., Schäuble, J., Babrowski, S. and Fichtner, W., (2013). User acceptance of electric vehicles in the French-German transnational context. in Proceedings of the 13th WCTR. Rio de Janeiro, Brazil.

GmbH, O. (2013). Stromtankstellen-21.de. Source: http://www.stromtankstellen-21.de/standort-suchen/ort/Karlsruhe (accessed on 05/06/2013)

ISST, F.-I. f.-u. (2010). Semantische Interoperabilität. Band 1: Eine Technische Einführung White Paper.

Lee, J., Park, G-L. and Kim, H-J. (2011). Reservation-Based Charging Service for Electric Vehicles. ICA3PP 2011 Workshops. Melbourne, Australia.

Sammer, G., Meth, D. and Gruber, CJ. (2008). Elektromobilität – Die Sicht der Nutzer. e&i elektrotechnik und informationstechnik, pp. 393–400.

Stadtraum. (2001). Einfach laden, Ladeinfrastruktur und Bezahlmodele für Elektrofahrzeuge. Flottenmanagement, pp. 68-70.

Standardization, E. C. (2011). Focus Group on European Electro-Mobility – Standardisation for read vehichles and associated infrastructure. CENELEC.

Wei H., Williard, N., Chen, C. and Pecht, M. (2013). State of charge estimation for electric vehicle batteries using unscented kalman filtering, Microelectronics Reliability. Microelectronics Reliability, Volume 53, Issue 6, pp. 840–847.

6 European harmonisation in the context of electromobility from a legal point of view

Eva Weis, Silvia Balaban

Center for Applied Legal Studies, Karlsruhe Institute of Technology

Attaining the government's objective of having one million EV on Germany's roads by 2020 (Bundesregierung 2009, p.2) and two million in France (Negre 2011, p.32) will not least depend on a possible cross-border use of the vehicles. Therefore, equivalent conditions concerning the infrastructure in Germany as well as in France are necessary for the use of EV.[21] Different than other new technologies the establishment of electromobility in Europe causes instantaneous harmonisation requirements. The significant difference is that EV have a mobility purpose, which does not only occur in a national context but also in an international environment. An interoperable use of EV across borders, requires unrestricted charging processes abroad. For that reason EVSE has firstly to be suitable to the physical requirements of the EV, such as plugs, second, the ICT-infrastructure has to be compatible (cf. chapter 1).

The main problem which occurs in the context of legal harmonisation of electromobility scenarios is that there is not even a national legislation, which could be used as prototype. Indeed, Germany for example launched an electromobility act not before 2015 (Bundesregierung 2014). However, the only contents regulated by this legislation are questions concerning road traffic regulations. Other urgent topics, such as an at least rudimental legal framework, about how requirements are set for EVSE are missing.

[21] The first approach of the European union to harmonize E-Mobility systems can be seen in the standardization mandate 468, which addresses the unification of charging systems for electric vehicles especially concerning plugs, see (European Commission 2010)

The electromobility law only deals with questions about priorising the use of EV within the traffic system.

The subsequent work firstly shows the framework conditions of the regarded situation. Afterwards the present chapter deals with general ways of attaining a European harmonisation within its limits and shows the harmonisation approach of the European Union (EU) on the example of fast charging systems. Lastly other aspects connected to the use of EV will be roughly outlined regarding their need of being harmonized.

6.1 Regarded situation

The present article deals only with public EVSE, since the European harmonisation issues occur because of travelling abroad and in this case such EVSE will primarily be used. The directive proposal COM 2013 (18) final[22] names public EVSE "publicly accessible recharging or refuelling point" and defines them in Art. 2, (5) as "a recharging or refuelling point which provides non-discriminatory access to the users". Another condition of the present examination is that it is assumed that the billing happens downstream towards charging process. It is intended, that customers receive only one bill from their own contractual partner (usually the domestic electricity provider) (Thomas et.al. 2010; Weis 2014, p. 43), even in cases of international charging processes. The possibility of prepaid or cash methods[23] of payment are not part of this examination. The background of only focusing on this payment method is the suitability of EV to act to the benefit of the local grid, whilst charging accordingly to the current electricity supply or even by returning energy into the grid (Rolink et.al. 2012, p. 87f; Schill 2010, p. 148ff; Timm & Vierbauch 2011, p. 71).

[22] The following article takes still the proposal version of the upcoming directive as basis, because the adopted version was not published at the time of finalizing this article.

[23] Payment methods like credit card or ec card are also understood hereunder.

6.2 General possibilities and limits of European harmonisation

At this point the general possibilities and limits of European harmonisation will be outlined in order to show how such a harmonisation is possible. The unification of systems could either be achieved by a simple agreement of the producers[24], like it has been done for mobile phone chargers in 2009, or by a technical norm, established by a standardization body and choosing only one system as state-of-the-art. Due to the fact, that both these ways of having a unified system are not legally binding, a European harmonisation has to be considered in order to enable a real incorporation in the legislation. As different legal systems are a barrier for a common European market a legal harmonisation might be the only valid resort in order to prevent a union-specific malfunctioning (Bieber et.al. 2012, sect.14 rec.2).

A legal harmonisation can be initiated for the establishment of a European market, in the meaning that there are no internal borders which ensure at the same time the free movement of goods, persons, services and capital. Art. 26 TFEU states further on that the Union shall adopt measures with the aim of establishing or ensuring the functioning of the internal market. According to Art. 114 TFEU the EU can use (inter alia) directives and regulations as legal instruments to achieve this aim. In contrast to the regulation which is directly valid the day it comes into force in all European countries at once, the directive needs to be implemented into national law by the member states.

Harmonisation measures can only occur if barriers to trade exist or lead to significant distortions of competition (Streinz 2008, rec. 925). A directive or a regulation must therefore not be used in an abusive way in order to attain harmonisation in an area where the EU has no legal competence. Art. 114

---

24 See: http://europa.eu/rapid/press-release_IP-10-1776_de.htm?locale=en; http://ec.europa.eu/enterprise/sectors/rtte/files/chargers/chargers_annex_i__list_of_signatories_en.pdf

TFEU should only be used for the purpose of avoiding the establishment of obstacles to trade due to a heterogeneous development of the national provisions. Such obstacles however need to be very probable and the measure must help to avoid it (Bieber et.al. 2012, sect.14 rec.13; EuGH, C-350/92, Slg. 1995, I-1985, rec. 35), as well as to improve the establishment and the functioning of an internal market (EuGH, Rs. C-66/04, SlG. 2005, S. I-10553, rec. 44). The mere finding of disparities and the abstract risk of obstacles to the exercise of fundamental freedoms or of distortion of competition are consequently not sufficient (EuGH, Rs. C-376/98, Slg. 2000, S. I-8419, rec. 84; Bieber et. al. 2012, sect. 14, rec. 13), they must be perceptible (EuGH, C-300/89, Slg. 1991, I-2867, rec. 23). Compliance with these principles is therefore a vital aspect for enabling a legal harmonisation.

However, unified provisions as a result of a legal harmonisation cannot be guaranteed without restrictions. If such a harmonisation would lead to an infraction of fundamental rights/ freedoms or the principle of subsidiarity/ proportionality, a harmonisation is not legally permissible. Especially reasons relating to public interest must override (Grabitz et.al. 2012, art.114, rec.53) and the balancing of interest between the union objective and a national provision must clearly lead to the necessity of having harmonized provisions (Dauses 2012, rec.148). This is in particular the case when national regulations are not sufficient.

6.3 Electromobility harmonisation and legal competences

Indeed, in the field of electromobility the European Commission identified the necessity of acting and therefore established a proposal for a directive on the deployment of alternative fuels infrastructure. In Art. 4 COM (2013) 18 final the directive outlines rudimentary provisions concerning electricity supply for transport. The most important already formerly discussed issue is the harmonisation of the plug systems used for electromobility, which shall be regulated in Annex III COM (2013) 18 final. The reason for the unification of plug systems is to enable a cross border use all over Europe.[25] At this point, the legal backgrounds of harmonisation will be outlined based on the example of DC fast charging systems (Mode 4), in order to show why the European Union can act in the field of electromobility and release such a provision leading to unified systems.

6.3.1 Example: The harmonisation of fast charging systems

The starting position of the discussion concerning plugs was the development of different plug systems. In the field of fast charging systems mainly the CCS and the CHAdeMO system competed. These different fast charging systems represent an obstacle to cross-border fast charge EV. A legal harmonisation is therefore required in the field of fast charging socket systems.

The European Commission bases the directive proposal COM (2013) 18 final on the enabling clause of Article 91 lit d) TFEU. According to Art. 90 and 91 TFEU, the Union has the legal competence to act in the field of transport. The Union initiative in this field is mandatory as the Member States do not have the necessary legal instruments to attain the necessary unified infrastructure design across Europe. European action can therefore provide the organisation of the infrastructure in a unified manner and the implementation of common standardized technical specifications for a single plug system. Legal harmonisation effectively removes the obstacles to trade which result from different national regulations and can enable a real option for the marketability of these products by setting up uniform standards. Hence, real internal market conditions are established (Möstl 2002, p. 321). As the limits of the European harmonisation, like the encroachment of fundamental rights, or fundamental freedoms as well as the principle of subsidiarity and proportionality are not infracted, the legal harmonisation is necessary for the plug systems and can be based on Art. 90 and 91 of the TFEU. The Regulation COM (2013) 18 final regulates that all EVSE in Europe should at least be equipped with a Type Combo 2 socket system and recommends the CCS for DC fast charging.

---

[25] Compare recital (30) COM (2013) 18 final.

6.3.2 Other possible legal competences of the EU in the electromobility context

Besides the enabling clause in the field of transport (Art. 90 and 91 TFEU) the European Commission (EC) also has the legal competence for environmental protection, which could also allow legal harmonisation concerning other aspects of unifying electromobility systems, which might not directly be related to transport competence. That would be, for instance, the case, if EV would prospectively be used to contribute to the stability of the electricity system, for example through feeding power back to the grid (so called vehicle-to-grid). In the moment when an EV is used for feeding power back, the original purpose of "transport" recedes. Therefore, it is questionable if such harmonisation issues could rather be based on the legal competence of environmental protection. As the EC is also seeing the protection of the environment as a European objective, a European harmonisation could also be possible based on the competence set down in Art. 191 in conjunction with Art. 192 TFEU. A legal harmonisation would therefore also help to foster environmental protection. This can be regulated in a more effective manner at European level as the environmental protection cannot be constrained by national boundaries. Hence, this might be worth to be considered for further regulations in the electromobility context.

6.3.3 Further problems and possible harmonisation issues of electromobility

Besides the necessity of harmonizing the charging socket system there might be other aspects in the context of electromobility which call for harmonisation. For these it has to be taken into account that a harmonisation could either be achieved through formal measures by the EC or through standardization in another way as mentioned in 6.2. To avoid misunderstandings it has to be said that harmonisation aspects mentioned in the following do not necessarily have to be regulated by formal legislation. However, if formal harmonisation measures are chosen the EC might have the possibility to justify its actions as being based on the legal competence in the field of transport and/or environmental protection.

At this point other exemplary aspects connected to the use of EV, which are either not or just rudimentarily covered by the already mentioned directive proposal COM (2013) 18 final will be roughly outlined regarding their need of unification, without indicating how unification should be achieved. The problems of harmonizing an international system for using EVSE can be seen from two different perspectives. On the one hand the problems can be viewed by the specific parts that a functioning electromobility infrastructure system will need. From this point of view infrastructure like identification systems or billing systems have to be mentioned.[26] The other perspective takes the different regulation areas into account. In the first place energy law, as a special area of competition law, has to be mentioned. Besides that especially measurement law can have an impact on the establishment of a European harmonized electromobility market. An overview over both perspectives shall now be considered.

6.3.4 Harmonisation issues from an infrastructure perspective

Example 1: Identification system

When thinking about harmonisation topics, the identification system should be considered. Concerning this matter both the technical questions about how to subscribe at a public EVSE and a standard for the structure of the customer IDs, if applied, have to be taken into account, so that every user, vehicle or contract can be identified throughout Europe. Meanwhile, the standardization of the identification number scheme is making progress. ISO 15118-2 now contains a regulation which allows identification throughout Europe. The other mentioned issues of how a user can sub-

[26] The same would apply for the already examined plugs.

scribe at a public EVSE are still under discussions. Indeed, ISO 15118-1 also contains regulations concerning this issue, but still offers different possibilities. The currently discussed options are e.g. plug & charge (authentication via the cable connection EV-EVSE), credit card payment and using other communication infrastructures, such as mobile phones, local wifi, and to directly register at a public EVSE (Fluhr & Stich 2013, p.290f). Two discussed identification mechanisms are Plug & Charge and the use of RFID cards. Indeed both of the systems can be used simultaneously, but the specific EVSE has to offer the respectively used mechanism. Without preferring one of the discussed mechanisms the question has to be asked if it is economically reasonable to implement both systems at public charging spots or if it would be better to agree on one of the methods. One problem of harmonizing this issue is that there is no agreement in for example German national context either (ISO 15118-1, p.23f.). But this missing national regulation could also be a chance to avoid sunken investments and agree early to a European standard.

Example 2: Billing system

Another exemplary part of an electromobility system for cross border charging processes is the billing system, which probably needs to be harmonized at least in a rudimentary way. Regarding this it has to be noticed that there are currently two different concepts of how metering data, which are the basis of the accounting, can be collected. In case of mobile metering[27] the electricity meter is either assembled within the EV itself or integrated into the power cord. Other than this, the concept of stationary metering locates the electricity meter within the EVSE. Both concepts have advantages, but also suffer from some disadvantages. Indeed, this harmonisation issue seems to be solved since Art. 4 No. 6 COM 2013 (18) final rules that "all publicly accessible recharging points for EV shall be equipped with intelligent metering systems as defined in Article 2(28) of Directive 2012/27/EU [...]", but this regulation does not necessarily lead to

[27] See for this system for example (Hechtfischer & Pawlitschek 2012; von Hammerstein & von Hoff 2011; Berg et. al 2011; Hechtfischer & Pawlitschek 2011).

the conclusion that only the stationary system would be legitimated. This provision could also be understood in a way, that just the total meter, which also measures the internal consumption of the EVSE, has to be an intelligent metering system integrated into the recharging point. If the regulation is understood in this way, there is still a need to harmonize the used system in Europe. A mixture of those two systems will not lead to adequate results, because the mobile system, at least if it is integrated directly in the EV, will measure all charging processes regardless whether there is another measuring device for a single charging process in the EVSE and also irrespectively whether the charging process takes place in a national or international environment. As a consequence, processes allowing the clearing of the different measurements have to be implemented for national as well as for international issues.

Another issue concerning the billing system is that the supplier of the customer needs to receive the measurement value to invoice the charging process. In Germany, for example, there are dedicated processes enforced by the federal network agency, which define how meter data has to be handled (Bundesnetzagentur 2006, p.37ff; Bundesnetzagentur 2010, p.63ff). For international charging processes it is necessary to harmonize a further process concerning the handling of meter data, because it is expected that the billing of both the contractual partner towards the customer and the billing between the involved companies will be based on those values.[28] In conclusion, there has to be a harmonized way on how required data incurring by charging processes can be exchanged. Otherwise there is the danger that the data exchange processes differ between the various member states or even between the different operators of EVSE and have therefore to be implemented in many different ways by a single company. It is obvious that this is not economically reasonable.

28 Indeed today there are also billing concepts based for example on time of use tariffs, but if electric vehicles will be used for grid beneficial purposes there has to be an invoice based on real-time measuring values. Another billing concept is in this case hardly imaginable, see also (Weis 2014, p. 42f).

6.3.5 Harmonisation issues from a law sector perspective

Besides the infrastructural perspective possible harmonisation issues can also be considered from a law sector perspective. Indeed, not all fields of law with effect on the establishing of EVSE have impact with regard on international charging processes. For example, the law of public streets and roads influencing the establishment of EVSE is irrelevant for international contexts, since it only concerns the question whether EVSE may be built on a specific place. Fields of law with impact on international charging processes are for example Energy and Measurement Law.

Energy law

First of all, in the context of electromobility, Energy law has to be taken into account. National legal frameworks dealing with energy law, like in Germany the EnWG, are already today affected by European legislation. For example the electricity directive 2009/72/EC or the energy efficiency directive 2012/27/EC influences the national energy legislation.

The basic question concerning electromobility in the context of energy legislation is, if either EV have to be seen as part of a smart grid and therefore as part of the energy market, or if they are just regular consuming appliances and the energy market ends with the EVSE. Due to the fact that EV shall also be used to act to the benefit of the grid (Rolink et.al. 2012, p. 87f, Babrowski et al., 2014), prospectively a perspective which includes EV into the energy market will probably be more effective by designing a future intelligent energy system. Following this reflection, the legal framework has to adopt electromobility issues besides the classical stationary market, since EV and also EVSE are part of this strictly regulated market.

Another fundamental question and one of the main topics currently discussed in Germany is a related issue: It is argued whether there is a right to access public EVSE through other suppliers than the provider itself or not. The legal background of this discussion is based on the question if the

EVSE is part of the grid and therefore a regulated infrastructure.[29] This is not only problematical because of the national legal positions but also because the regulation of the grid is based on European Law. Especially the right to access grid infrastructures is a fundamental condition of European legislation. Because of that a single member state cannot change its regulation if the European perspective tends to see the EVSE as part of the grid as well. The proposal of the upcoming directive about an alternative fuels infrastructure is not clear regarding this point. In fact, recital 14 COM 2013 (18) final formulates that "publicly accessible recharging points are currently not part of the regulated activities of a distribution system operator as defined in Chapter VI of Directive 2009/72/EC". On the other hand the directive proposal COM 2013 (18) final also cites in Art. 4 No. 8 that "Member States shall not prohibit EV users from buying electricity from any electricity supplier regardless of the Member State in which the supplier is registered". Independently of the conclusions drawn by the directive proposal, the question has to be answered immediately since it determines many further legal questions related to public EVSE in national contexts as well as in international cases.

Due to that fact and because of the already existing European legislation regarding the stationary electricity market, the EC is asked to take up position regarding these issues with view of international charging processes. Without legal certainty even in the national contexts, it is hardly possible to develop a system concerning cross-border cases. In the future an approach could be the consideration of roaming system in the telecommunication sector to see if there are processes which can be carried over to the international electromobility sector.

[29] See about this dispute (Weis 2014, p. 194ff; Hartwig 2013, p. 478ff; von Hoff 2009, Lüdemann et. al. 2014, p. 6; Feller et. al. 2010; Fest et. al. 2010, p. 95f; Michaels et. al. 2011, p. 831; Böwing 2013, p. 97; Heinlein 3013, p. 192ff; Boesche 2013, pp. 201f).

Measurement law

Also the measurement law is partially determined by European legislation. The European Measurement Instruments Directive (MID) 2014/32/EU[30] regulates amongst others technical requirements that active electrical energy meters have to fulfil. Since public EVSE shall be equipped with intelligent metering systems[31] and a measurement of energy by selling electricity to electromobility users is in view of the fact that EV can prospectively be used to the benefit of the grid (Rolink et.al. 2012, p. 87f) and also because of the necessary efficiency goals[32], the regulation concerning energy meters has to be kept in mind. Consideration should be given to develop electromobility specific requirements, since the current specifications in MID cannot be transformed without problems in all regards (Weis 2014, p.75ff). Since energy meters for public EVSE will be used all over the European Union the reasons leading to enact the MID also exist in this case. Instead there has been already an attempt to incorporate EV chargers into the scope of the MID (European Commission 2011, p.21ff), but it failed because the approach was not based on measuring topics but rather on the harmonisation of plugs.

6.4 Conclusion and outlook

European harmonisation is an important and precious instrument for avoiding heterogeneous developments within the Member States. Especially in the field of electromobility diverging regulations are counterproductive. Therefore, unifying systems all over Europe is mandatory. The harmonisation of the charging socket system is indeed a very valuable step towards a cross border use of EV. Nonetheless, as seen, the directive proposal COM (2013) 18 final does not capture all relevant issues in the context of electromobility, which means that there are still some topics left open, which might as well be worthwhile to be harmonized. In regard to

[30] The directive 2014/32/EU replaced the as MID formerly known directive 2004/22/EC.
[31] Art. 4 Nr. 6 COM 2013 (18) final.
[32] See directive 2012/27/EU.

the other mentioned issues it would now be useful to start a discussion about whether and in what manner unification is needed. However, there is currently no need to overregulate a slowly starting market. Still the basic frame conditions should be determined and so far as possible already in a harmonized way enabling international charging processes. In so doing the upcoming electromobility market gives, due to the possibility of an early unification, a real chance to avoid mistakes which were for example made with regard to other technologies in the past. Since electromobility systems are not yet well established in the Member States, there is an opportunity to build an interoperable system from the bottom up, without major changes in national regulations. In this way, financial risks could also be reduced. Nevertheless the starting point needs to be set up currently.

6.5 References

Babrowski, S.; Heinrichs, H.; Jochem, P. and Fichtner, W. (2014): Load shift potential by electric vehicles in Europe, Journal of Power Sources 255, pp. 283-293

Berg, A. & Zayer, P. & Hechtfischer, K. & Pawlitschek, F. (2011). On-board Metering – einfach überall Strom tanken. Energiewirtschaftliche Tagesfragen (ET) number 6, 58-61.

Bieber, R., & Epiney, A., & Haag, M. (2012). Die europäische Union. Baden-Baden.

Boesche, K. (2013). Einordnung der Ladeinfrastruktur als Kundenanlage im Sinne des § 3 Nr. 24a EnWG. In: Boesche, K. & Franz, O. & Fest, C. & Gaul, A. (ed.), Berliner Handbuch zur Elektromobilität, München, pp. 201-207.

Böwing, A. (2013). Umsetzung der Elektromobilität im Umfeld des heutigen Elektrizitätsmarktes. In: Boesche, K. & Franz, O. & Fest, C. & Gaul, A. (ed.), Berliner Handbuch zur Elektromobilität, München, pp. 87-99.

Bundesnetzagentur (2006) Anlage zum Beschluss BK6-06-009 vom 11.07.2006 in der Fassung der letzten Änderung durch den Beschluss BK6-11-150 vom 28.10.2011.

Bundesnetzagentur (2010) Anlage 1 zum Beschluss BK7-09-001 vom 09.09.2010 in der Fassung der letzten Änderung durch den Beschluss BK6-11-150 vom 28.10.2011.

Bundesregierung (2009). Nationaler Entwicklungsplan Elektromobilität der Bundesregierung. online via http://www.bmub.bund.de/fileadmin/bmu-import/files/pdfs/allgemein/application/pdf/nep_09_bmu_bf.pdf [last accessed 31.08.2014]

Bundesregierung (2014). Referentenentwurf zum Entwurf eines Gesetzes zur Bevorrechtigung der Verwendung von elektrisch betriebenen Fahrzeugen, processing state 07/29/14.

Dauses, M. (2012). EU-Wirtschaftsrecht. München.

European Commission (2011). Report on the Measuring Instruments Directive 2004/22/EC, Background document 4 – Details relevant for impact assessment of 13 sectors. online via http://ec.europa.eu/DocsRoom/documents/6587/attachments/1/translations/en/renditions/native

European Commission (2010). Standardisation mandate to CEN, Cenelec and ETSI concerning the charging of electric vehicles. online via http://www.etsi.org/images/files/ECMandates/m468electricalvehicles.pdf

Feller, D. & de Wyl, C. & Missling, S. (2010). Ladestationen für Elektromobilität – regulierter Netzbereich oder Wettbewerb. Zeitschrift für Neues Energierecht (ZNER) 240-146.

Fest, C. & Franz, O. & Haas, G. (2010). Energiewirtschaftliche und energiewirtschaftsrechtliche Fragen der Elektromobilität – Teil 1. Energiewirtschaftliche Tagesfragen (ET), number 4, 93-99.

Fluhr J. & Stich (2013), V. Identifikation in der Elektromobilität. In: Boesche, K. & Franz, O. & Fest, C. & Gaul, A. (ed.), Berliner Handbuch zur Elektromobilität, München, pp. 286-296.

Grabitz, E., & Hilf, M., & Nettesheim, M. (2012). Das Recht der europäischen Union. München.

von Hammerstein, C. & von Hoff, S. (2011). Neue Infrastrukturkonzepte für Elektromobilität. Zeitschrift für Neues Energierecht (ZNER) 259-264.

Hartwig, M. (2013). Öffentliche Ladestationen als Teil des Elektrizitätsversorgungsnetzes der allgemeinen Versorgung – Teil 2. Zeitschrift für Neues Energierecht (ZNER) 475-482.

Hechtfischer, K. & Pawlitschek, F. (2011). Das Auto lernt zählen: Mobile Metering – Ladeinfrastruktur für E-Fahrzeuge. Automobil Elektronik number 02, 58-60.

Hechtfischer, K. & Pawlitschek, F. (2012). Mobile Metering: Auf mobiler Messtechnik beruhende Ladeinfrastruktur, Infrastrukturrecht (IR) 278-280.

Heinlein, B. (2013). Einordnung der Ladesäuleninfrastruktur als Energieversorgungsnetz. In: Boesche, K. & Franz, O. & Fest, C. & Gaul, A. (ed.), Berliner Handbuch zur Elektromobilität, München, pp. 191-200.

Lüdemann, V. & Jürgens, C. & Ortmann, M. (2014). Datenschutz und Elektromobilität: Zum Ladevorgang an öffentlichen Ladestationen. Recht der Datenverarbeitung (RDV) 3-10.

Michaels, S. & de Wyl, C. & Ringwald, R. (2011). Rechtsprobleme im Zusammenhang mit der Nutzung des öffentlichen Straßenraums für Elektromobilitätsanlagen. Die öffentliche Verwaltung (DÖV) 831-840.

Möstl, M. (2002). Grenzen der Rechtsangleichung im europäischen Binnenmarkt – Kompetenzielle, grundfreiheitliche und grundrechtliche Schranken des Gemeinschaftsgesetzgebers. EuR, 318-350.

Negre, L. (2011). Livre Vert sur les infrastructures de recharge ouvertes au public pour les véhicules «décarbonés»: http://www.developpement-durable.gouv.fr/IMG/pdf/Livre_vert_L-_NEGRE_Infrastrucutres_recharge_pour_les_vehicules_decarbones.pdf

Rolink, J. & Hoerenkamp, W. & Rehtanz, C. (2012). Ladeinfrastrukturen für die Netzintegration von Elektrofahrzeugen, Automatisierungstechnik (at) 84-91.

Schill, W.-P. (2010). Elektromobilität in Deutschland – Chancen, Barrieren und Auswirkungen auf das Elektrizitätssystem, DIW – Vierteljahreshefte zur Wirtschaftsforschung number 79, 2, 139-159.

Streinz, R. (2008). Europarecht. Heidelberg.

Thomas, I. & Weidlich, A. & Johns, M. (2010). IT-Gestützte Geschäftsprozesse in zukünftigen E-Mobility Szenarien, in: VDE-Proceedings, VDE-Kongress – E-Mobility: Technologien – Infrastruktur – Märkte, Paper FKV 7.3.3, Berlin.

Timm, C. & Vierbauch, C. (2011). Elektromobilität – Chancen durch wertschöpfende Geschäftsmodele nutzen. Energiewirtschaftliche Tagesfragen (ET) number 3, 68-71.

von Hoff, S. (2009). Zugangsanspruch zu Elektromobilitätstankstellen. Zeitschrift für Neues Energierecht (ZNER) 341-345.

Weis, E. (2014). E-Mobility – Energierechtliche Modelierung einer integrationsfähigen IKT-Infrastruktur für öffentliche Ladestationen. Baden-Baden.

7 Analysis of the utilization of the EV in CROME

Matthias Pfriem, Frank Gauterin, Thomas Meyer

Institute of Vehicle System Technology, Karlsruhe Institute of Technology

7.1 Data acquisition

The data acquisition in CROME is the basis for the technical analysis of the use of the EV. Therefore, information about trips and charging events has been collected using five different data acquisition systems by the OEM and KIT (cf. chapter 1). To cope with the heterogeneity of the data resulting from different samples, methods of acquisition and data logging devices has been a major challenge.

7.1.1 Definition of a common data sample

In a first organisational step, a common data sample was agreed between KIT as research institution and the OEMs. This negotiation had to find a satisfactory compromise between the interest of research to gain as much information as possible and the interest of the companies to protect their intellectual property, which lies e.g. within the operational strategy of the EV. The minimum requirement was a sample that enables for the analysis of key figures of mobility and the evaluation of the cross-border usage of the EV. Another central aspect was the identification of national idiosyncrasies in EV usage that can be found in the comparison of the sub-fleets of the two countries. In addition, for several vehicle types, further data has been gathered as the basis for a technically oriented analysis and a more precise investigation of the energy consumption for mobility and comfort needs. Table 11 gives an overview of the common data sample, which was agreed on, including expansions for additional analysis.

Table 11: Data acquisition during trips and charging events.

	Trips	Charging Events
Date and time	X	X
Vehicle position (longitude, latitude, altitude)	X	X
Indoor and outdoor temperature	X	X
Utilisation of heating and air conditioning	X	
Remaining range	X	
Vehicle speed	X	
Total mileage	X	
Actuation of brake and accelerator pedal	X	
Longitudinal and lateral acceleration	X	
SOC of battery	X	X
Engine revolution speed and engine torque	X	
Energy consumption		X

Depending on the dynamic of the signals, different groups of acquisition frequencies have been defined.

For some EV types data could merely be collected in histograms due to technical constraints of the vehicle data acquisition systems. Furthermore, the systems were unable of collecting GPS-data, which was seen as crucial for the analysis of cross-border usage. Thus, an additional measurement system has been developed to fill this gap.

7.2 Smartphones as data loggers

The need for the completion of the acquired trip data of two EV types motivated for the development of a custom-built measurement system. Its main data acquisition task was seen in the collection of GPS data and its automatic transfer via mobile radio communication. Moreover, it should be easily applicable to the EV as a retrofit with preferably no need for a fixed installation, as the EV had already been distributed to customers at this time.

Smartphones with a dedicated data logging application had been identified as the ideal choice for those requirements (cf. Pfriem et al., 2014). They are capable of collecting GPS data with position, speed and altitude as well as accelerations. The connection to the mobile cellular network enables for data transfer after each trip, what was the basis for parallel data analysis throughout the active phase of the fleet test. In addition, this regular read-out of data is the basis of an operation unlimited in time as the memory can be cleared after each transfer of trip data and a monitoring of user's activity is ensured (i.e. sending reminders for using the application is possible). A major advantage over traditional automotive data logging equipment is the touch-screen as a comfortable interface for user interaction. Driver input was used to identify the driver via unique ID to be able to differentiate between individual persons in multi-user scenarios. Furthermore, an optional multiple-choice questionnaire has been implemented to collect additional data about the trip (if the user didn't decide to skip it). This additional data consists of the number of adults and minors on board, the trip purpose, the payload, the intention for a cross-border trip and the combination with other means of transport during the trip.

Figure 35: User interface of data logger application

A post-processing method was developed for the exploitation of the acceleration data from the smartphones. This procedure serves the purpose to recalibrate the axis-system of the acceleration data, based on the measured data. It is necessary, as the data is recorded in an internal axis system of the smartphone, which is mounted in the car for each trip in a suction mount in an unspecified position with an unknown rotation to the vehicles axis system. The acceleration data is needed for further analysis in the vehicle coordinate system according to DIN 70000. Therefore, three Euler rotations from the source axis-system to the target axis-system have to be defined. The first two rotations are defined during the phases in which the EV is without movement, where gravity is measured as the only acceleration. This defines the orientation relative to the vertical axis of the vehicle as the axis is in opposite direction to gravity. Slope mistakes are minimized by the evaluation of all stop-phases of the trip, where in mean there should be no resulting slope angle. The longitudinal axis of the EV is identified by the evaluation of acceleration events from stop phases, as they occur at intersections or traffic lights. The direction of the acceleration during straight-forward acceleration events defines the longitudinal axis of the car and the angle for the final rotation. A control of the straightness of the acceleration event is done with an evaluation of the GPS bearing signal. The lateral axis is defined as rectangular to both other axes and therefore does not need a separate calibration. In an additional rotation the cross-talk between longitudinal and lateral acceleration is minimized to remove imprecisions in the axis recalibration process, which might have remained up to this point. Further information about the method to use smartphones as data loggers has been published in (Pfriem and Gauterin, 2014).

7.3 Key figures of mobility

The basic analysis of mobility in the fleet test is founded on data of about 120,000 trips from six different vehicle types. They covered a total length of about 600,000 km in a driving time of approximately 20,000 hours.

7.3.1 Trip length

The trip lengths show a variation between the different vehicle types from a median of 15 km for the Porsche Panamera Plug-In Hybrid to just above 2 km for the Mercedes-Benz A-Klasse E-Cell (see Figure 36).

The Porsche Panamera data has an interquartile range of over 40 km with an upper quartile of 45 km and shows a clear difference in usage in comparison to the others. This might be a result of the plug-in hybrid topology of the Panamera while all other cars are BEV. Another difference is the vehicle segment, where the Panamera stands apart as a luxury sports limousine. It will be left out of the group for further analysis, which will be focussed on BEV. The group of BEV shows, that the distances driven are mostly far below the vehicles' maximum range of at least 135 km. The car with the longest range shows the lowest distance median and the two with the shortest maximum ranges rank second and third. This demonstrates that the driving distances are mainly a result of the individual usage scenarios and not of the vehicles' characteristics.

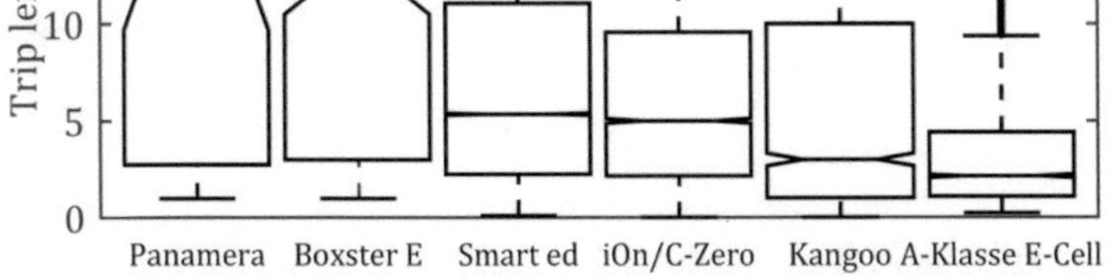

Figure 36: Boxplots for comparison of trip lenghts per vehicle type.

The combined analysis of BEV trip lengths reveals a clear focus on short-distance mobility with a mean length of merely 8.3 km per trip and a median of 5.1 km (see Figure 37).

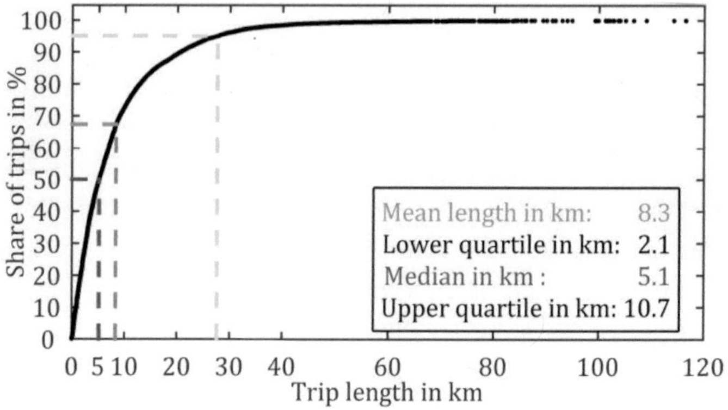

Figure 37: Empirical distribution function of BEV trip lenghts.

This is significantly shorter than the average trip length in Germany of 11.5 km (Infas et al., 2010). Only 5 % of the trips account for distances beyond 28 km.

7.3.2 Trip duration

The predominant short trip lengths result in mainly short durations with more than half of all trips shorter than 15 minutes (see Figure 38). The median is 13.1 minutes and the mean value lies at 20 minutes. This is again shorter than the average trip in Germany with 23.2 minutes (Infas, 2010). Only 19 % of all trips last longer than 30 minutes.

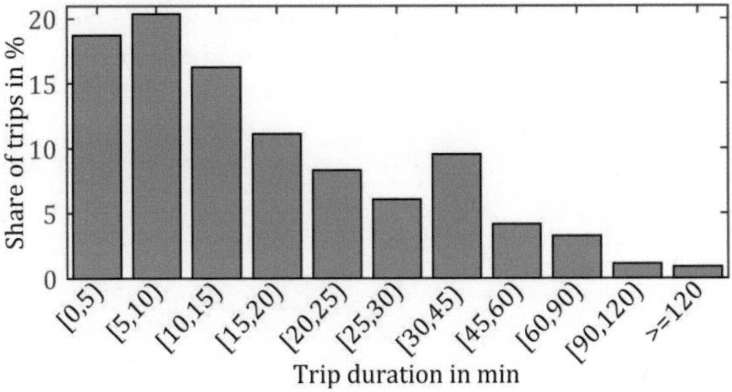

Figure 38: Classing of trip durations.

7.3.3 Mobile days

A mobile day is defined for this analysis as a day on which the EV has been employed for at least one trip. The combined analysis of trip lengths on mobile days reveals that the cars are used for an average distance of 28.3 km and a median of 20.3 km per mobile day. Again, this is far below the maximum range of the vehicles and even 95 % of all mobile days stay below 80 km, which is only 60 % of the shortest maximum driving distance of the BEV.

It has been found by Franke (2013) that EV users tend to prefer a range buffer of an average of 20-25 %. Even with 25 % range buffer and the smallest maximum range of the EV, over 98 % of all mobile days can be covered. And with recharging during the day this percentage can be raised to almost 100. The data likewise reveals that in rare occurrences the EV are employed beyond their maximum range, as they are recharged during the day between trips.

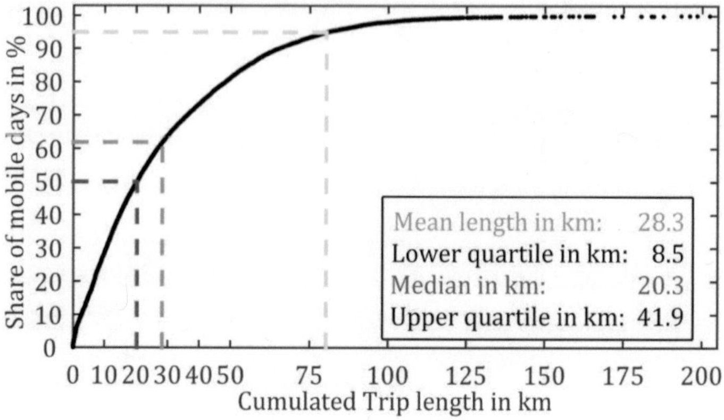

Figure 39: Empirical distribution function of combined trip lengths on daily basis.

7.3.4 Movement area

The GPS data shows, that the EV are mostly used in a constrained local area. The movement area is defined as the smallest circle that includes all GPS values measured per car. In this analysis, EV with at least 50 trips with valid GPS data are included. The mean radius of 32.7 km is strongly influenced of one outlier with a movement radius almost twice as large as the second largest radius.

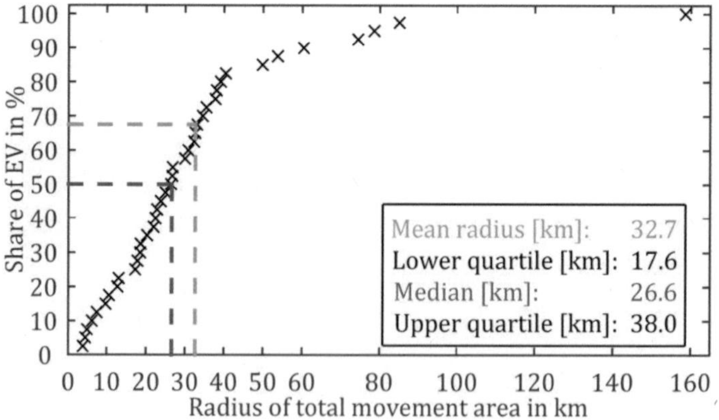

Figure 40: Distribution of the radius of the total movement area.

Half of all vehicles are employed in an area with a radius of less than 26.6 km and therefore never conduct long-distance trips.

7.3.5 Battery capacity usage

The analysis of the usage of the battery capacity again shows the battery to be overdimensioned for most cases in the monitored usage. About three quarters of all trips start with a SOC above 60 % and the median lies at 79 % (see Figure 41). The consumption per trip has a median of 4.5 % and an upper quartile of ca 10 %. In result, the distribution of end-SOC has a median of 71 % and a lower quartile of still 53 %.

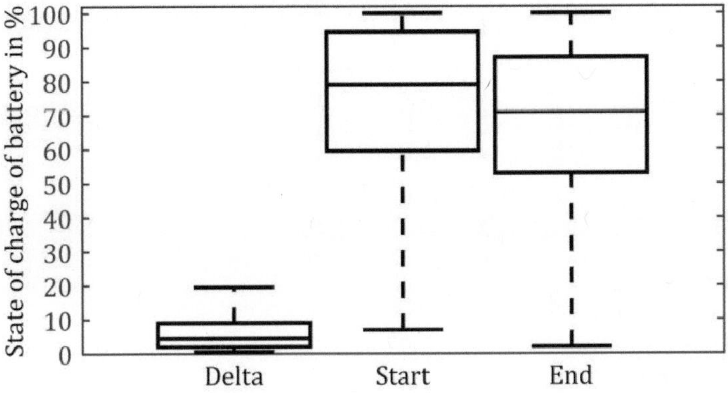

Figure 41: Boxplots of SOC difference per trip and start and end values.

7.3.6 Cross-border mobility

One project goal was to create a framework for cross-border mobility including charging. Likewise, the cross-border mobility was monitored via GPS tracks. Crossings of the French-German-border have been monitored for several vehicles with both countries as starting point. In sum, about 17,500 trips could be analysed in this aspect, though much of the trip data does not include GPS information. The analysis showed a share of 0.7 % of the trips to be cross-border. This should not be seen as a result of possible limitations in cross-border charging. It can be seen as the effect of the strong regionally limited focus of the vehicle users, which has already been pointed out.

7.4 Charging analysis

The analysis of the charging behaviour reveals that the batteries of the EV are overdimensioned for most cases in the fleet test.

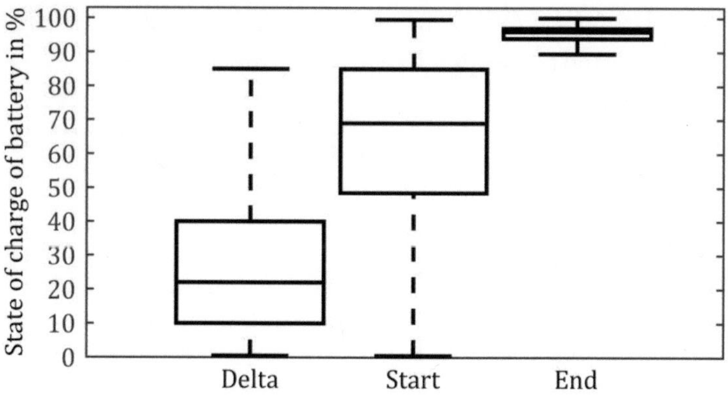

Figure 42: Boxplots of SOC difference per charge and start and end values.

The median of the SOC at the start of the charging process is 69 % and three quarters of the charging processes start above 48.5 %. In general, the EV are fully charged with few variation of end-SOC. Likewise, there is usually only little energy transported with a median of 22 % SOC per charge. This habit offers a huge potential for bidirectional charging as the vehicles could usually serve as a source of electricity at the beginning of the charging process with more than half of their total battery capacity. On the other hand it offers the potential for a significant cost reduction by employing smaller battery sizes, which fit the demand of energy for mobility better.

The charging durations show a median of 119 minutes and stay far below the plug-in time.

Furthermore, it has to be noted, that the charging speed decreases towards high SOC (see Figure 44) and therefore the charging processes seem rather slow, as they usually happen with a large share in high SOC classes.

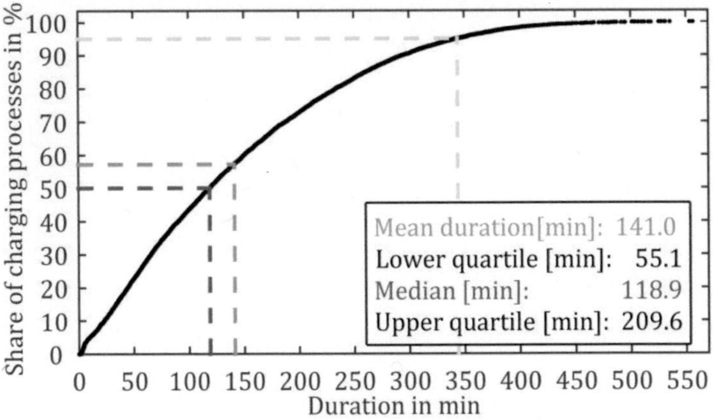

Figure 43: Empirical distribution function of charging duration.

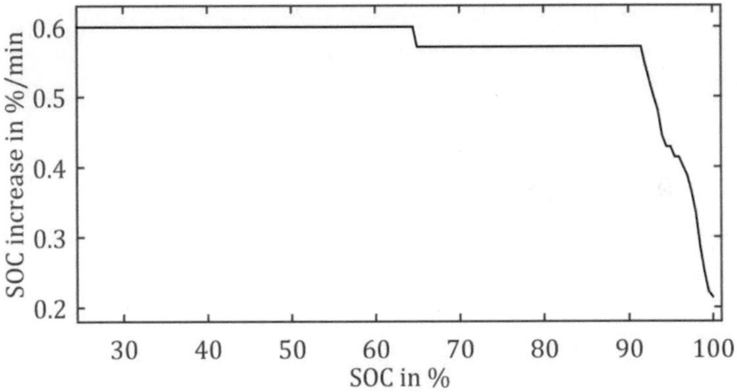

Figure 44: BEV charging speed at different SOC levels.

The EV are recharged very frequently with a median of one day, despite the fact that they are not used on every day. Therefore, the number of recharging processes per mobile day is even higher. As a result of the low depth of discharge, it can be expected, that the batteries will show only small effects of wear and tear through the usage.

7.5 Comparison of Germany and France

The comparative analysis of the mobility between Germany and France is based on the evaluation of Smart ed data to eliminate distortive factors like different ranges or top speeds of the cars, as other EV models were (mainly) solely available in one of the countries. 52 technically identical EV build the base for the comparison and are almost evenly distributed between both countries. The total tracked trip length is more than 490,000 km in aggregated data and over 50,000 km in GPS-traces at 1 Hz acquisition rate. Further information can also be found in (Pfriem and Gauterin, 2014).

7.5.1 Trip length

The comparison of the trips in both countries shows a stronger use of the EV in Germany for distances between 1 and 15 km while EV used in France are driven comparably more often for trips in distance classes between 15 and 50 km (see Figure 45).

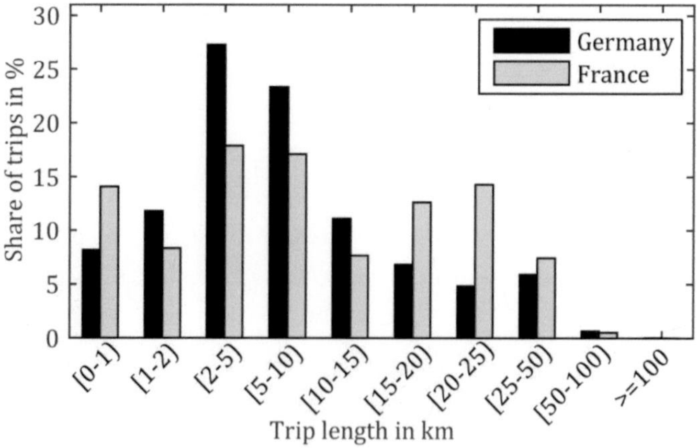

Figure 45: Trip length distribution in Germany and France.

The biggest difference can be found for trips between 20 and 25 km, which are three times as likely to happen in France as in Germany. Long trips beyond 50 km are very uncommon in both countries and we can in general observe a focus on trips below 10 km length.

7.5.2 Trip length on mobile days

The EV are used for combined ranges up to 200 km per day. With an official range of 135 km in the New European Driving Cycle (Smart, 2010), the EV must have been recharged significantly during the day for those cases. But these occurrences are very rare in both countries and only about each 2 % of mobile days show distances beyond 100 km (see Figure 46).

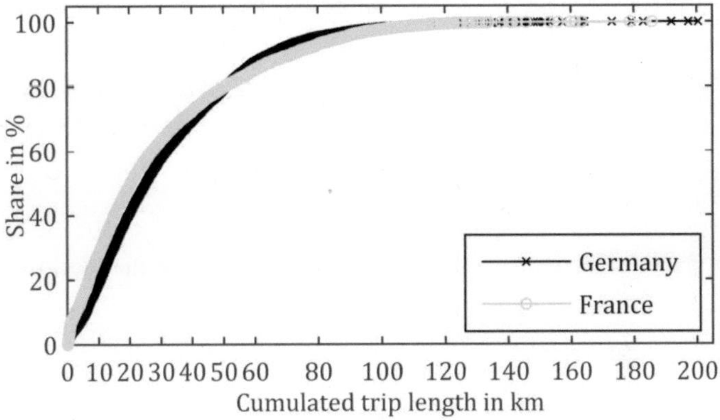

Figure 46: Empirical distribution function of combined trip lengths on daily basis in Germany and France.

Despite the trend to shorter distances on individual trips, the EV used in Germany are driven on longer cumulated daily distances than the EV used in France. In the fleet test, the median of 25.2 km in Germany is about 25 % higher than in France with a median of 20.4 km per day as a result of a slightly higher number of trips on mobile days in Germany than in France.

7.5.3 Trip duration

In the fleet test, German EV trips last significantly longer than French trips in terms of time with a median of 14.4 min, which is about 20 % higher than the French median of 12.1 min (see Figure 47).

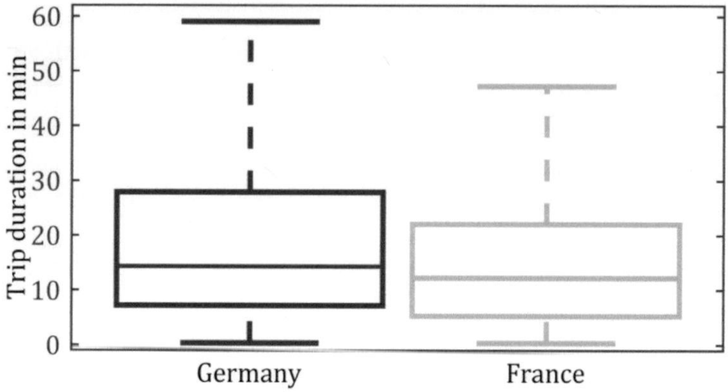

Figure 47: Boxplot of trip durations.

At first glance, this result seems contradictory to the distribution of trip lengths. But the explanation for that can be found in the comparison of the speed profiles of the trips.

7.5.4 Speed profiles

The median of the driving speed for German trips is 24 km/h, whereas the French median of 28 km/h is approx. 17 % higher, what allows for longer trips in shorter time. The variation in mean speeds between trips is almost identical in both countries.

The speed profiles for both countries show, that the EV are predominantly used at urban speeds below 60 km/h (see Figure 48). Both curves also present an unusual peak at the top speed of 100 km/h to which the analysed EV type is limited. This demonstrates that users regularly experience speed limitation, which could be a possible obstacle for acceptance. The data suggests that a higher top speed would suit the users' needs better. The curve of the German drivers shows peaks at 30 km/h, 50 km/h and at 70 km/h, what are common speed limits in Germany, whereas the French speed distribution shows a much higher share around 90 km/h, the common speed limit for rural roads in France. The most significant difference is the higher share of top speed in France, what corresponds to the higher percentage of long trips. Long trips usually take place with a significantly bigger portion of high speeds than short trips.

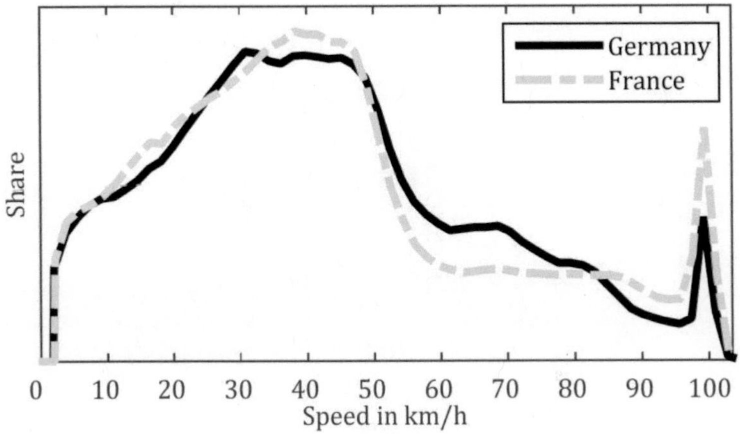

Figure 48: Probability density function of driving speed.

7.5.5 Mobility in the course of the day

The use of the EV throughout the day shows clear differences between the countries (see Figure 49). The French use is distributed more or less symmetrically around the middle of the day, whereas the German usage reveals a focus on the first half of the day. Another difference is, that the French mobility is shifted one hour towards the afternoon in comparison. This may be due to the fact that German workers tend to start and stop working earlier than the French, which can also be observed in the comparison of departure times in large German (Infas, 2010) and French (Commisariat général au développement durable, 2011) Mobility studies.

This result is of special importance for the integration into the electricity grid. The EV are very likely to be connected to the grid at the end of the working day after the last trip for charging. In case of immediate charging, the peak load on the grid of both countries will occur at different times as a result of different national working cultures. From a European perspective, this could offer a potential for cross-national grid stabilisation in the future.

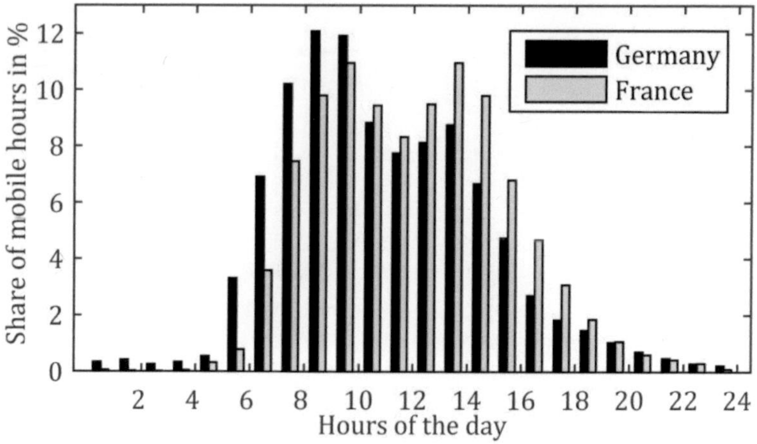

Figure 49: Probability density function for hours with mobility.

7.6 Conclusion and discussion

Within the CROME project we were able to gather mobility data from approximately 120,000 trips with EV. We merged data from pre-installed CAN data loggers and from a newly developed measurement system of a smartphone with a custom-built application in the fleet vehicles.

It was shown, that the vehicles are predominantly used on short trips in an urban environment within a limited geographical area. Most of the EV were employed within small corporate fleets, where there was no need for long EV trips due to a regional focus of business or the alternative of conventional cars for long distances.

It was also pointed out that there are some notable national idiosyncrasies in our fleet test. Individual trips on the French side are longer in terms of distance, whereas the combined trip lengths on a mobile day are longer on the German side. The French trips show a faster speed profile with a larger share in speed regions for rural roads, resulting in a higher average speed. This is an effect of the less dense settlement structure in the French project region. The analysis showed that the EV are already suitable for the usage in corporate fleets. In most cases, a maximum of half of the battery capacity was used between two charge processes and range showed to be sufficient with a notable buffer for almost all mobile days.

However, the results from the analysis of only approximately 100 EV cannot be transferred directly to all business situations. Although, many companies and especially regional authorities show comparable usage profiles to the project's test drivers. Especially small and medium-sized businesses often have a regional business focus. In Germany 80 % of all commercial vehicles are used less than 80 km per day (Hacker, et al., 2011). Therefore, many companies could as well benefit from a share of EV or a total replacement of vehicles with combustion engines within their fleet in terms of energy efficiency and the mitigation of CO_2 emissions (cf. Ketelaer et al. 2014). From a cost perspective, for regionally focussed businesses it would be better to choose EV with a smaller, demand-based maximum range to avoid an overinvestment into the expensive battery, when it is almost never used to full capacity.

A larger share of EV in corporate fleets would presumably result in a corresponding offer of second-hand EV for private customers in a few years. Currently more than 60 % of all new vehicle are registered by commercial owners (Kraftfahrt-Bundesamt, 2014). Therefore, it is important to build EV that suit the needs of corporate users to address a large share of the vehicle market. This could be a significant support of the market ramp-up of BEV and affect the offer for private users within a few years as well.

7.7 References

Commisariat général au développement durable, "Les véhicules électriques en perspective: analyse coûts-avantages et demande potentielle", 2011.

Franke, T., "Nachhaltige Mobilität mit begrenzten Ressourcen: Erleben und Verhalten im Umgang mit der Reichweite von Elektrofahrzeugen", Chemnitz, 2013.

Hacker, F., Harthahn, R., Herrmann, H., Kasten, P., Loreck, C., Seebach, D., Zimmer, W., "Betrachtung der Umweltentlastungspotenziale durch den verstärkten Einsatz von kleinen, batterieelektrischen Fahrzeugen im Rahmen des Projekts „E-Mobility“", Berlin, 2011.

Infas and DLR, "Mobilität in Deutschland 2008. Ergebnisbericht, Struktur – Aufkommen – Emissionen – Trends", 2010.

Kraftfahrt-Bundesamt, http://www.kba.de/DE/Statistik/Fahrzeuge/Neuzulassungen/Halter/2013/2013_n_halter_dusl_absolut.html?nn=1174638, 2014.

M. Pfriem, F. G., "Employing Smartphones as a Low-Cost Multi Sensor Platform in a Field Operational Test with Electric Vehicles", 47th Hawaii International Conference on System Sciences (HICSS) (pp. 1143-1152). Waikoloa, Hawaii, USA: IEEE, 2014.

M. Pfriem, F. G., "Selected Results from a Large-Scale Field Operational Test with Electric Vehicles in Germany and France", 5th Hybrid an Electric Vehicle Conference. London, GB, 2014.

Smart, Smart fortwo electric drive, 2010.

8 Adoption of EV in the French-German context

Axel Ensslen, Patrick Jochem, Martin Rometsch, Wolf Fichtner

French-German Institute for Environmental Research /
Institute for Industrial Production, Karlsruhe Institute of Technology

A significant data source for the user evaluation has been created during the three online interviews during the project phase. Between September 2013 and July 2014 answers to the third survey on acceptance of EV used in a professional environment within the CROME project were collected. This survey focused on the users' and fleet managers'[33] long-term experiences with EV in the French-German project region. Professional as well as private EV adoption intentions are analysed in this article. After presenting a brief literature review, this chapter characterises the sample of CROME's third online questionnaire. The survey participants' sociodemographic background, their mobility behaviour as well as EV usage patterns are described in detail. Furthermore, the participating organisations, their fleet managers and the vehicle fleets within their organisations are characterised. Afterwards, relevance of different car purchase decision criteria in the private and organisational context are compared. As EV users' expectations given in the first two surveys concerning range, purchase price and charging time are only satisfied to low degrees (cf. Schäuble et al. forthcoming and Ensslen et al. 2015), these barriers are analysed in detail by measuring the EV users' willingness to pay, the minimum range required and acceptable charging times. Furthermore, the participants' stated preferences concerning their next car purchase intentions are analysed and evaluated. A binary logistic regression model is derived from the survey data providing information on private EV purchase intentions. This model takes into account socio-economic data, mobility behaviour, survey partic-

[33] The persons in charge of the EV in the companies who have partly been involved in the decision making process to acquire the EV.

ipants' attitudes towards the environment as well as their willingness to seek for information on future developments of EV in order to characterise potential future EV adopters. The sample is suitable for research questions on private EV purchase intentions, as (i) the sample is comparably unbiased, i.e. the major part of the survey participants had not bought an EV for their private purposes when they completed the survey, so self-selection process might be negligible (cf. Globisch et al. 2013). Furthermore, (ii) the professional EV users had already experienced EV, which is a decisive factor influencing EV purchase intentions (cf. Ensslen et al. forthcoming). The purpose of the chapter is to analyse purchase intentions of a specific population: EV users and fleet managers who experienced EV during a period of about two years in a professional context.

8.1 The third survey in the context of the CROME project

Comparing market developments of EV[34] in France and Germany leads to the conclusion that the French market for EV currently rather captures the position of a lead market, as in France 37,100 EV were registered in June 2014 whereas only 24,000 EV were registered in Germany in July 2014 (cf. Nationale Plattform Elektromobilität 2014). The French car market is considerably smaller than the German (cf. Dudenhöffer et al. 2014). Germany on the other hand is rather characterised as lead supplier for EV together with the United States (cf. Nationale Plattform Elektromobilität 2014). The Nationale Plattform Elektromobilität (2014) even provides an explanation for the high dynamics of EV diffusion in countries like Denmark, France, the Netherlands and Norway. According to the authors the monetary incentives provided constitute a large lever for EV diffusion (cf. Pfahl et al. 2013). The Nationale Plattform Elektromobilität (2014) compares the markets from a macroeconomic point of view by considering national framework conditions, national EV stock and number of charging points.

34 Within this study 99 % of the respondents were users of pure electric vehicles (BEV). In this article BEV users are considered and referred to as EV users.

Furthermore, information on German EV users' willingness to pay (WTP) for EV and their perception of the EV specific particularities concerning range and charging infrastructure are provided (cf. Nationale Plattform Elektromobilität 2014 and Plötz et al. 2013).

In order to analyse German and French EV users' and fleet managers' perception of EV, user acceptance has been analysed within the CROME project based on a multi-methodological, interdisciplinary approach (cf. Figure 50). The acceptance analysis as part of the evaluation concept consisted of repeatedly questioning the CROME participants using and managing the EV in a professional environment by online surveys with different focuses (expectations, first experiences and long-term EV adoption), as well as of face-to-face interviews of some private and professional users (cf. chapter 4) and workshops with fleet managers. Additionally, technical data on trips such as speed and acceleration were collected by using data loggers and smartphones (cf. chapter 2).

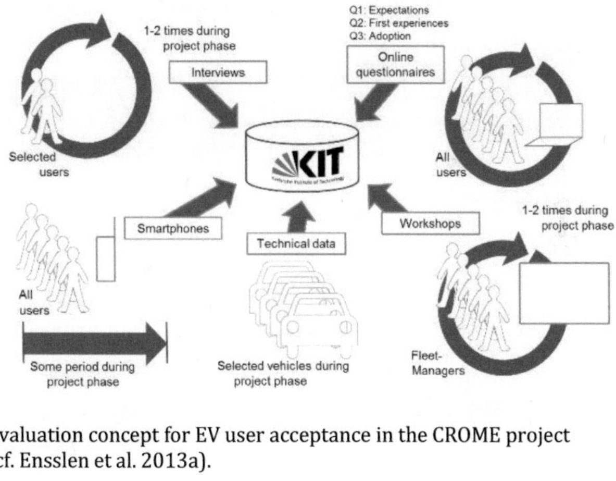

Figure 50: Evaluation concept for EV user acceptance in the CROME project (cf. Ensslen et al. 2013a).

Before the project CROME started in 2011, the projects Kléber in Strasbourg (cf. Pierre 2014; Pierre 2015) and MeRegioMobil (Paetz et al. 2012) in the region of Karlsruhe analysed EV user acceptance in the Upper Rhine region, but from a rather national perspective. Afterwards the projects

CROME, ELEC'TRA (cf. Tanguy et al. 2015) and RheinMobil (Stella et al. 2015) analysed EV user acceptance from a binational point of view.

General information on CROME can be found in Gagnol et al. (2013). Details on the subjective and qualitative part of the user acceptance study, i.e. results from the interviews with professional and private EV users and fleet managers are available in Pierre (2014). Details on current research on technical issues of EV, i.e. the analyses of data from dataloggers in the vehicles are available in Pfriem and Gauterin (2013), Pfriem et al. (2014) and chapter 7.

This section puts its focuses on the findings of the online questionnaires among the professional EV users within the multi-methodological, interdisciplinary approach of the CROME project. More precisely we are focusing on the findings of the subjective and quantitative part of the user acceptance analysis, i.e. the results of the first two online questionnaires on the CROME participants' expectations (Q1) and their first experiences (Q2) are presented before corresponding research questions for the third online survey (Q3) and this article are derived. Furthermore, early private EV adopters amongst a specific population are characterised, notably amongst EV users who experienced EV during a period of about two years in a professional context.

8.1.1 EV users' expectations (Q1)

Between 2011 and 2013 the car manufacturers participating in the CROME project (cf. CROME 2014) could convince some organisations purchasing EV to participate in the CROME project and to take part in the EV user acceptance studies. In order to collect quantitative data on the EV users' and fleet managers' opinions on EV usage and to learn about their expectations, the first online survey was distributed directly to the fleet vehicle users after their organisations decided to take part in the CROME project. Furthermore, the survey was completed by addressing the participating organisations' fleet managers and decision makers. In order to analyse EV users' and fleet managers' expectations responses to the first survey were collected from September 2011 until April 2012 (cf. Ensslen et al. 2012). Most of the EV users stated that the EV are allocated in their employers' vehicle fleets and are used by several persons, predominantly for professional trips. At that point of time some of the EV users had already made first experiences with the EV, others did not have any experience with the EV at all. Ensslen et al. (2012) tackled the research question whether there are differences concerning user acceptance of EV between the French and German respondents. According to their results, the French respondents seem to be more optimistic about purchasing an EV for their private purposes within the next 10 years than the German EV users. Less of the French answered "maybe" and more of the French chose the answer "yes". The question, who of the EV users within CROME could envision purchasing an EV for private purposes within the next years was further analysed in order to identify main characteristics of early EV adopters in the French-German context. Therefore, Ensslen et al. (2015) developed two binary logistic regression models. One of the two models estimated the EV users' and fleet managers' purchase intentions based on their socio-economic background, their experience levels with EV as well as their car usage behaviour. Results show that respondents who could envision purchasing an EV within the next 10 year are likely to have a higher level of income, to have a household equipped with two or more cars and to travel more than 50 kilometres a day on average, not necessarily by car. This model additionally shows that possibilities to experience EV (e.g. by test drives) are important leverages to support adoption of EV by private car buyers. As organisations kept on joining the CROME project until August 2013, 238 responses to the first questionnaire were collected from September 2011 until November 2013 (cf. Figure 51).

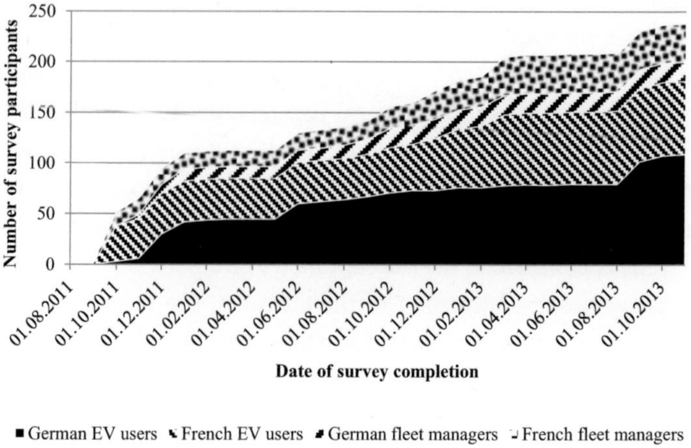

Figure 51: Date of survey completion of the first survey (Q1) in CROME.

8.1.2 EV users' and fleet managers' first experiences (Q2)

Based on the second survey which was completed between September 2012 and May 2013, the CROME participants are characterised according to their attitudes, values and norms. Ensslen et al. (2013b) characterised the survey participants, who had on average experienced EV for one year, based on 19 items assessing their individual innovativeness, environmental awareness, price sensitivity, attitudes to EV and the perceived image of EV. The second binary logistic regression model developed in Ensslen et al. (2015) used this characterisation and joined these factors with the respondents' answers provided in the first survey (Q1) on their EV purchase intentions. According to the model presented dependencies between the respondents' attitude towards EV, their innovativeness as well as the perceived image of EV can be observed. Dependencies between the respondents' EV purchase intentions and their environmental awareness as well as their price sensitivities could not be observed. Unfortunately these findings are limited as the model had to be estimated based on only 60 answers of respondents, who completed the first as well as the second questionnaire.

8.1.3 Research questions, methods used and structure of this article

Based on the findings of the two first online questionnaires (Q1 and Q2) presented above and results of the qualitative user acceptance study (cf. CROME 2014: pages 63 – 67) the third online questionnaire (Q3) was developed focusing on the professional EV users' and fleet managers' long-term experiences and their willingness to further adopt EV in the professional as well as the private context.[35] EV acceptance is assessed by individuals' EV purchase intentions. Therefore the field test participants were interviewed about their planned next car purchase profoundly. The article analyses the differences between criteria influencing purchase decisions in the private and organisational context. Statistical significance levels of differences are determined by nonparametric Mann-Whitney U tests[36]. Identifying respondents who have a higher probability to choose an EV when being asked about their next car purchase decisions is of particular interest. Hence, the focus is on describing powertrain choices based on exogenous variables accounting for attitudes, sociodemographic background and mobility behaviour. Corresponding dependencies are estimated by using binary logistic regression analysis.

In order to characterise potential EV adopters and EV specific barriers this article has the following structure: Section 8.2 presents the data sample including the sociodemographic background of the survey participants, their mobility behaviour, EV usage patterns and their employment by industry. Section 8.3 compares car purchase decision criteria mentioned by EV users buying cars for private purposes and fleet managers buying cars for professional reasons. Section 8.4 describes the professional EV users' next private car purchase intentions before section 8.5 assesses EV specific barriers provided by the EV users. Potential early EV adopters are characterised in section 8.6 before a conclusion and an outlook complete this chapter.

---

35 With further focus on additional services (cf. Chapter 3).

36 With IBM SPSS Statistics®

8.2 Sample description

The following analyses are derived from the third online survey (Q3) that was distributed to the professional EV users within CROME. This survey was completed by the EV users and fleet managers between August 2013 and July 2014 and focused on their long-term experiences with EV. Overall 134 responses from 67 French and German respondents were collected. 82 respondents stated to be only EV user, 37 stated to be fleet manager and EV user, 7 stated being fleet manager only and 5 provided the answer of neither being EV user nor fleet manager. 3 persons did not provide any answer in this issue. The EV used by the survey participants are owned by organisations in different sectors (cf. Figure 54).

8.3 Sociodemographic background

The majority of the survey participants are male (81 % of the Germans and 69 % of the French). On average they are about 45 years old (SD=10) and well educated, as more than half of the respondents have an academic background. Only 18 % of the French and 10 % of the German respondents live in single person households. More than a third of the survey participants did not want to provide information about their households' net income level. On average the German EV users have a net household income of 4,088 Euros and the French of 3,547 Euros (cf. Figure 52). Compared with national averages female participants are underrepresented. Well-educated male respondents living in multiple person households with a high income and being between 30 and 59 years old are overrepresented (cf. Ensslen et al. 2013a). Although this repartition is not representative for the French or the German population, it might be representative for the people working in the participating organisations.

The French as well as the German EV users participating in this survey majorly live in municipalities with less than 20,000 citizens whereas most of them state that their workplaces are located in municipalities with more than 20,000 citizens (cf. Figure 53).

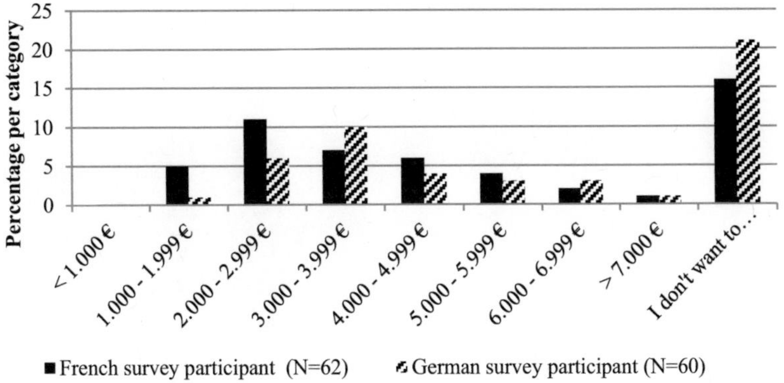

Figure 52: CROME EV users' monthly net household income.

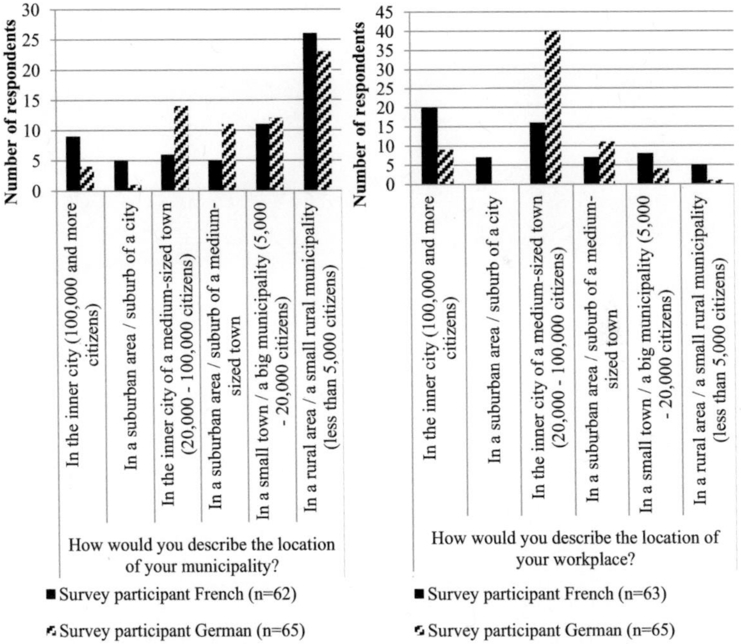

Figure 53: Number of citizens in the home municipalities of the EV users (l.h.s.) and in the municipalities of their workplaces (r.h.s.).

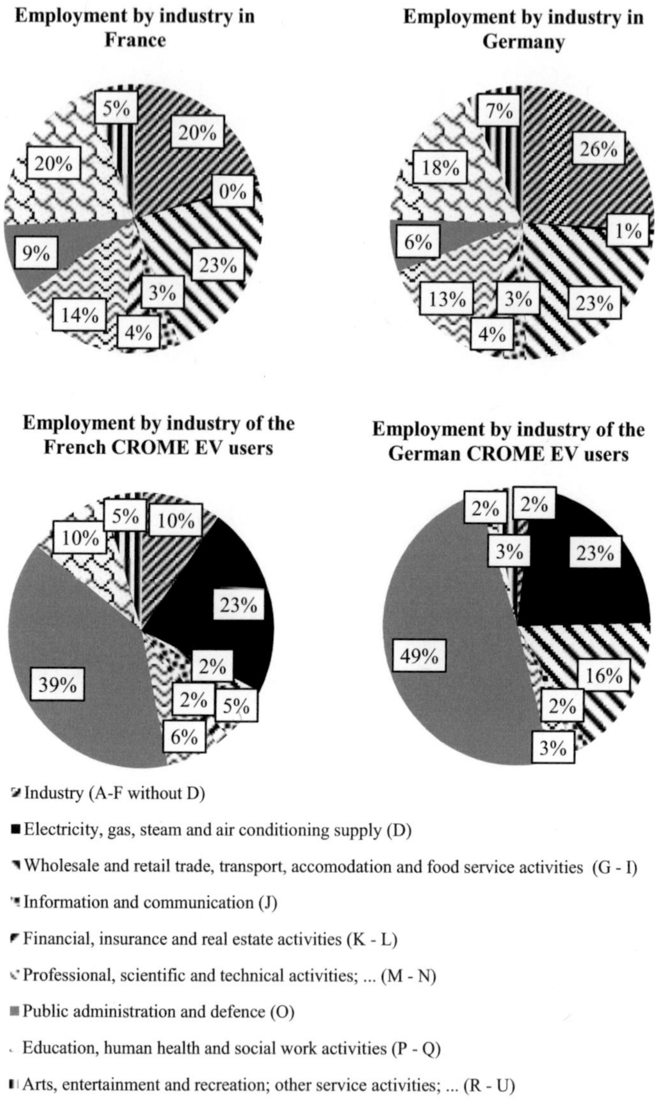

Figure 54: Employment by industry of the CROME participants (n=123) and employment by industry in France and Germany; Data source: Eurostat (2012).

44 % of the CROME EV users participating in the third online questionnaire state to be working in public administration. 23 % of them in the energy sector. Those two sectors are heavily overrepresented, shown by the distribution of employees in France and Germany (cf. Figure 54), as employment in these sectors normally makes up for less than 10 % of the total employment in France as well as in Germany. Only about a third of the survey participants work in other sectors representing an employment share of more than 90 % in France and Germany (cf. Figure 54).

The overrepresentation of the respondents in the sectors public administration and energy supply might be linked to these organisations' interests as e.g. development of new business segments. This might have been a driving factor for energy supply companies to participate. As several municipal utilities participated in CROME, they might have influenced public authorities to participate. Furthermore, employment stakes and the positive communication effect of showing to be environmentally aware and innovative by using EV might have been another important factor influencing organisations in the public administration sector to participate.

8.4 The respondents' mobility behaviour

Three out of four survey participants (n=117) use cars on a daily basis and about 20 % of them at 1-3 days per week. As the EV has predominantly been acquired as a fleet vehicle used on average by more than 10 persons for professional trips (cf. Table 12), it is used less frequently than cars in general (cf. Figure 55). 23 % of the survey participants use an EV on a daily basis, 34 % at 1-3 days per week and 25 % at about 1-3 days per month. About 18 % use an EV less frequently. The German field test participants use the EV less frequently on a highly significant level (cf. Figure 55) what could explain why individual trips of the German users are shorter but have longer cumulated daily distances than the EV used in France (cf. chapter 7.5.2). The fact that the EV used by the German survey participants are used by more people than the French EV further supports this assumption (cf. Table 12). Furthermore, about half of the survey participants state to do walking trips on a daily basis. The other modes of transport are less important for most of the EV users (cf. Figure 55). This might be linked to the fact that the major share of survey participants lives in rather rural areas, where public transportation is rare, and uses cars for commuting to the workplaces allocated in the inner cities.

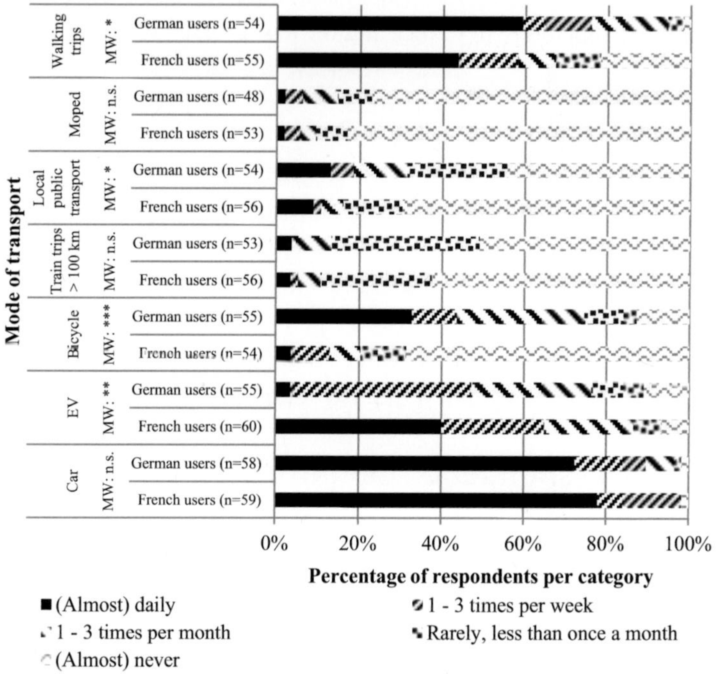

Mann-Whitney Test results (MW):
°p<0.1; *p<0.05 ; **p<0.01; ***p<0.001

Figure 55: CROME EV users' mode of transport.

8.5 EV usage patterns in CROME

As already stated, on average the French as well as the German survey participants experienced the EV for almost two years. The EV within the

CROME project driven by the French participants are used by less persons than the EV driven by the German field test participants (cf. Table 12).

Table 12: Experience, number of EV users and EV usage purposes in CROME.

			French EV users		**German EV users**		**P-Value (Mann-Whitney U Test)**
			N	**Mean**	**N**	**Mean**	
	Experience with EV (in months)		71	23.31	68	21.72	0.35
	How many persons use this EV?		47	11.91	21	12.95	**0.08**
I use the EV...	...to commute...	...during ... % of the trips.	69	12.78	56	4.64	**0.001**
	...for private purposes...		69	3.91	56	1.30	**0.003**
	...for transportation of goods...		69	7.39	56	0.00	**0.098**
	...for provision of services...		69	51.78	56	39.20	**0.099**
	...for transportation of persons...		69	10.72	56	2.71	**<0.001**
	...for other professional job completions...		69	13.41	56	52.14	**0.006**

Furthermore, the EV users were asked which purposes they use the EV for. It can be observed, that the EV usage scenario of the French survey participants is more diversified. The EV is used for commuting more frequently. It is also used more frequently for private purposes at week-ends or during holidays, for transportation purposes of goods (e.g. in order to transport products, material or machines) or in order to transport persons. The French as well as the Germans stated to use the EV very frequently for provision of services (e.g. installation, repairing, consulting, visiting and care); the German fleet test participants use the EV almost exclusively for professional trips (cf. Table 12).

8.6 Characterisation of the participating organisations and fleet managers

The fleet managers were asked about their influence on car purchase decisions in their company[37]. Most of the fleet managers do not take the car purchase decisions in their companies alone. Most of the respondents state to have a certain influence on their organisations' car purchase decisions (cf. Figure 56). This is the reason why this population is particularly interesting for providing information on future car purchase intentions, particularly EV.

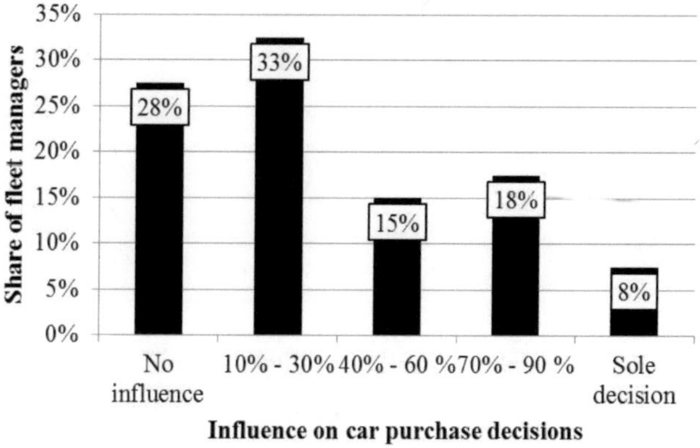

Figure 56: Fleet managers' estimate on their influence on future car purchase decisions within their organisations (n= 40).

The fleet managers were asked about the powertrains of the vehicles in their current fleet today and their estimation for 2020. As only fleet managers of EV owning organisations were questioned, alternative fuel vehicles including hybrids in general already account for 21 % of the participating organisations' vehicle fleets today (15 % EV, cf. Figure 57). According to the participating fleet managers the share of alternative fuel vehicles will be

[37] Differences between the French and German fleet managers' answers are not significant. N=40; U=192.5; Z=-0.206; p=0.837.

more than doubled until 2020 and PEV, i.e. PHEV, REEV, BEV will make up for more than 30 % of the vehicles in their organisations' fleets in 2020.

CROME might have been an important project for the participating organisations in order to experience future developments. According to the fleet managers the EV share in the participating French organisations is currently significantly higher than in the participating German organisations (cf. Figure 57). According to the fleet managers' answers the share of EV in 2020 will be significantly higher in the French organisations (cf. Figure 58).

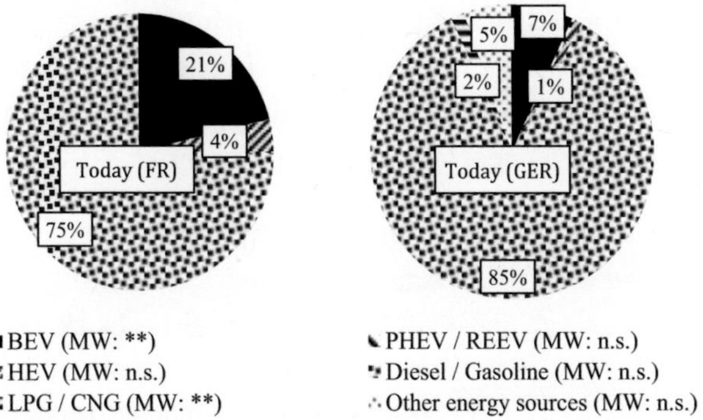

Figure 57: French (l.h.s; n=25) and German (r.h.s.; n=18) fleet managers' descriptions for the composition of their current organisations' vehicle fleets.

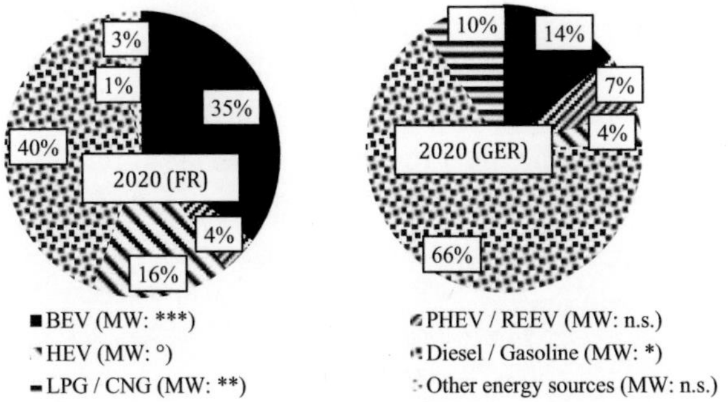

Figure 58: French (l.h.s; n=25) and German (r.h.s.; n=18) fleet managers' descriptions for the composition of their current organisations' vehicle fleets.

8.7 Car purchase decision criteria – private vs. professional

The fleet managers and the EV users within the CROME project were asked which criteria they consider during the vehicle buying process. Fleet managers were asked to provide information about the differences in their purchasing decision for private and business purposes. According to the results provided in Figure 60, price is the most important criterion for a large part of EV users and fleet managers. Surprisingly, the difference between the two groups was not significant for this criterion. According to the results provided in Figure 59 the next cars' emissions are of higher relevance in professional decision-making. On the other hand design, acceleration and fuel efficiency is more important for private car purchase decisions.

Engine size[38] and car brand[39] is more important to the French fleet managers, whereas versatility[40] is significantly more important to the German fleet managers. This might be linked to the differences observed concerning sectoral distributions of the French and German organisations participating in CROME and the purpose the fleet managers intend to purchase the cars for (e.g. vehicles as fleet cars or as company cars). The purchase decision criteria safety[41], emissions[42], versatility[43] and fuel efficiency[44] are more important to the German EV users than to the French.

Comparing the results of Figure 59 with the results of Peters & de Haan (2006) who asked 1307 Swiss households in representative telephone interviews in 2005 about the criteria influencing their decisions when purchasing new cars, the most important criteria (purchase price, car size, safety) are on average ordered in the same way. Differences can be observed concerning the ranking of the criterion fuel type. The findings pre-

38 $N=32$; Mann-Whitney $U = 68.5$; $Z=-2.178$; $p=0.029$.
39 $N=32$; Mann-Whitney $U = 76.5$; $Z=-1.870$; $p=0.062$.
40 $N=32$; Mann-Whitney $U = 66.5$; $Z=-2.324$; $p=0.020$.
41 $N=104$; Mann-Whitney $U = 1005$; $Z=-2.324$; $p=0.020$.
42 $N=104$; Mann-Whitney $U = 1071$; $Z=-2.093$; $p=0.036$.
43 $N=104$; Mann-Whitney $U = 987$; $Z=-2.491$; $p=0.013$.
44 $N=104$; Mann-Whitney $U = 881.5$; $Z=-3.248$; $p=0.001$.

sented in Figure 59 oppose the results of Peters & de Haan (2006) where fuel type is ranked behind the criteria design and gearshift. According to Figure 59 fuel type is similarly relevant as fuel consumption, total cost of ownership (TCO) and car brand and definitely of higher relevance than gearshift. Availability of EV and the possibility to experience the new technology might have influenced the CROME users and fleet managers to give fuel type a higher weight than the Swiss households back in 2005.

8.8 EV users' next car purchase decisions

The CROME sample of professional EV users is suitable for research questions on private EV purchase intentions, as (i) the sample is unbiased, i.e. the major part of the survey participants had not bought an EV for their private purposes when they completed the survey. Consequently self-selection processes might be negligible (cf. Globisch et al. 2013). Furthermore, (ii) the professional EV users had already experienced EV, which is a decisive factor influencing EV purchase intentions (cf. Ensslen et al. 2015). The CROME EV users participating in the survey were asked about their experience levels of purchasing cars[45]. Only 7 % of them state not ever having purchased a car. 84 % of the 125 EV users answering this question state having purchased a car at least twice; 31 % of the 125 EV users even stated having purchased a car 6 times or even more frequently. The EV users were asked whether they plan to purchase a new car or rather a second-hand car for their private purposes. About one third of them stated planning to purchase a new car[46]. Furthermore, the EV users were asked whether the next car they intend to purchase will be an additional car or whether it will replace another car in the household[47]. Only 9 % of the 123 EV users providing an answer to this question stated that the next car will be an additional car or the first car ever purchased in the household. 64 %

45 Differences between the French and German survey participants' answers are not significant. N=125; X^2=1.623; df=3; p=0.654.

46 Differences between the French and German survey participants' answers are not significant. N=121; X^2=0.283; df=1; p=0.595.

47 Differences between the French and German survey participants' answers are not significant. N=123; X^2=8.002; df=5; p=0.156.

stated that the next car will replace another car in the household. 27 % of the respondents did not know which vehicle they intend to replace or if they want to purchase an additional car (cf. Figure 60).

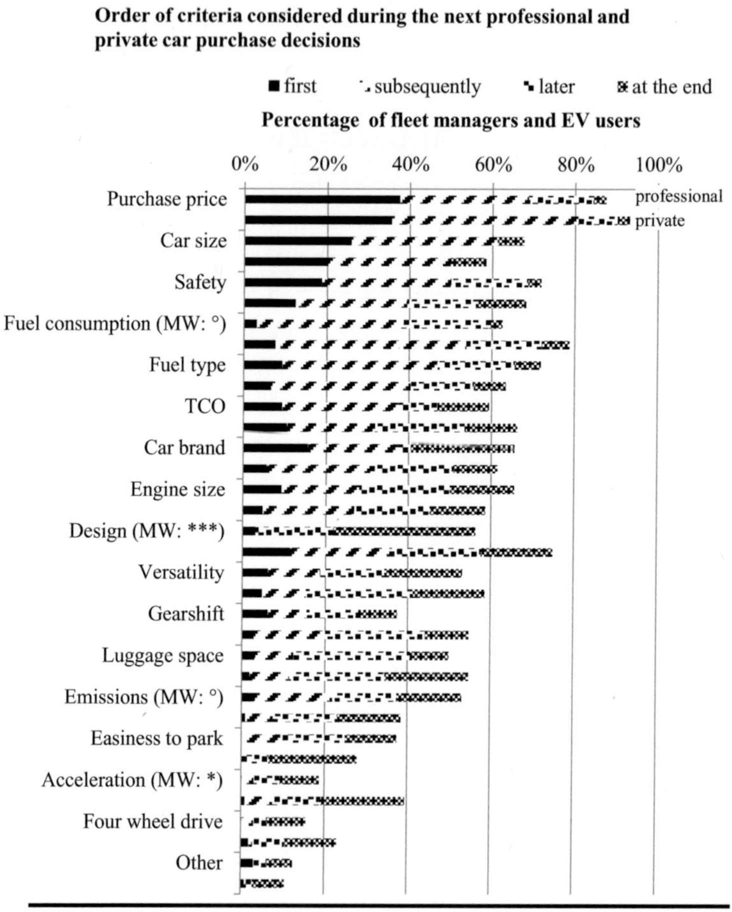

Mann-Whitney Test results (MW) for testing the difference between professional and private car purchase decisions (only calculated for selected criteria):
°p<0.1; *p<0.05; **p<0.01; ***p<0.001

Figure 59: Relevance of different criteria during the purchase decision of cars in the private and professional context (private n=104, professional: 32).

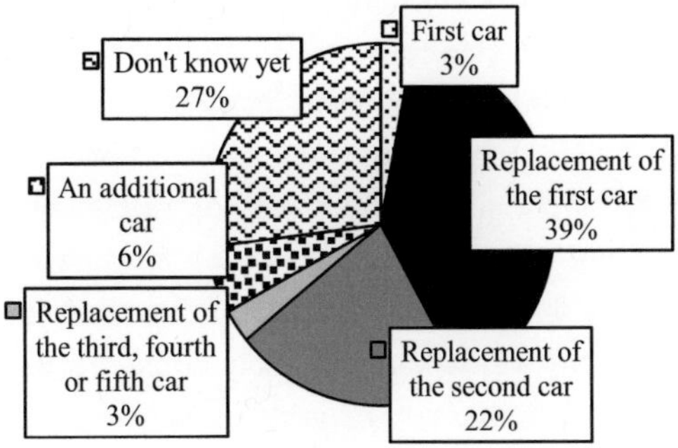

Figure 60: According to the respondents, will the next car be an additional car or a car replacing another car in the household? (n=123).

Being asked about medium and long-term car purchase intentions, 60 % of the EV users state that they will purchase a new car during the next 10 years[48]. 30 % of them state that they will maybe do so. Only 10 % state that they will not buy a car during the next 10 years (cf. Figure 61).

Figure 61: Statements of EV users on car purchase intentions for their private needs during the next 10 years (n=125).

The EV users were also asked whether they could envision purchasing an EV within the next 10 years for their private needs[49]. More than a fourth of them stated yes, about half of them was rather undecided and stated may-

[48] Differences between the French and German survey participants' answers are not significant. N=125; X^2=1.945; df=2; p=0.378.

[49] Differences between the French and German survey participants' answers are not significant. N=126; X^2=0.911; df=2; p=0.634.

be and one fifth of them stated not being willing to envision purchasing an EV within the next 10 years (cf. Figure 62). The same question had been asked in the first survey and was analysed for the CROME users providing an answer to this question between August 2011 and April 2012 (cf. Ensslen et al. 2012). Surprisingly the share of respondents answering "Maybe" increased somewhat and the share of respondents answering "Yes" decreased slightly, but not significantly[50]. The share of respondents providing the answer "No" remained relatively constant. Please note that only a small share of the survey participants of the third survey had also participated in the first survey, so the population in the samples changed. Therefore analysis concerning the development of the same users' private EV purchase intentions over time is not possible.

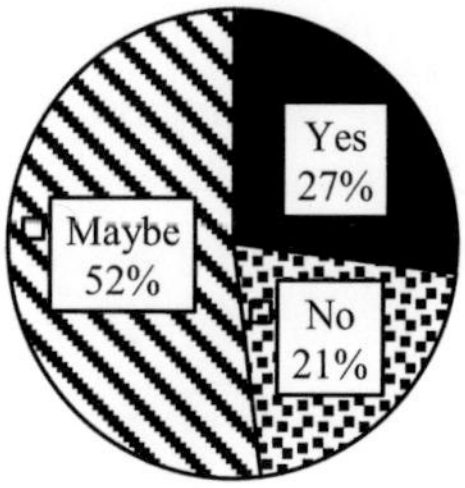

Figure 62: Could the EV users envision purchasing an EV within the next 10 years (n=126).

Furthermore, the EV users were asked to choose among different powertrain technologies when they buy their next car[51]. The respondents had the possibility to choose between the following alternatives: (1) internal combustion engine vehicles (ICEV) driven with gasoline, (2) ICEV driven with diesel, (3) liquefied petroleum gas (LPG) or compressed natural gas (CNG) vehicles, (4) BEV, (5) PHEV or REEV[52], (6) hybrid EV which can not be charged with electricity and (7) indifferent. About half of the 113 respondents chose ICEV, 14 % stated that they are indifferent concerning power-

[50] $N_{Q1,\text{April }2012}=108$; $N_{Q3}=126$; Mann-Whitney U = 6316.5; Z=-1.027; p=0.304.

[51] Differences between the French and German survey participants' answers are not significant. N=113; $X^2=9.792$; df=6; p=0.134.

[52] PHEV and REEV have been grouped together, as both powertrain technologies combine the possibility to charge the EV with electrictricity and to extend the vehicles' range by using gasoline.

train choice. Only 12 % chose BEV, but 28 % stated to choose PEV consisting of BEV, REEV and PHEV (cf. Figure 63). This is consistent with the 27 % of respondents stating that they could envision purchasing an EV within the next 10 years for their private needs provided in Figure 63.

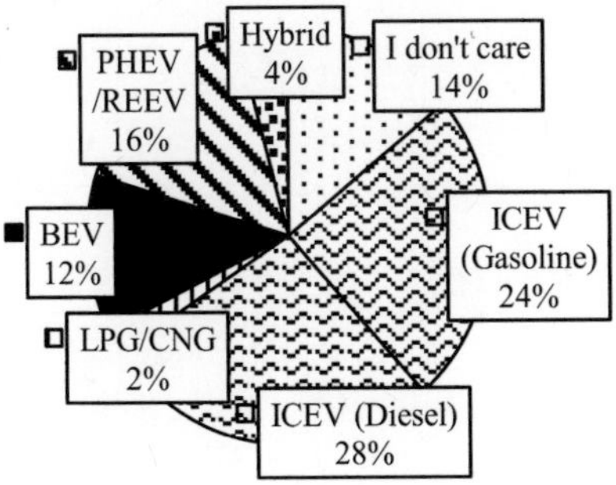

Figure 63: Which powertrain technologies state the EV users to choose when they purchase their next car? (n=113).

8.9 EV specific barriers

As no significant differences between the French and German respondents' answers concerning the minimum range of an EV can be observed[53], the cumulative distribution of answers provided by the 116 French and German EV users is presented in Figure 64. About 20 % of the EV users would be satisfied with a minimum range of 176 km, 40 % with 240 km and 70 % with 300 km. 90 % of the EV users would be satisfied with a minimum range of 500 km (arithmetic average: 308.28 km, median: 300 km).

The French and German EV users' expectations concerning charging time for 100 km range differ on a low significant level[54]. 50 % of the French respondents expect a charging time of 30 minutes or less whereas 50 % of the German respondents expect a charging time of an hour or less. Overall,

53 N=116; Mann-Whitney U = 1647; Z=-0.185; p=0.854.
54 N=116; Mann-Whitney U = 1362; Z=-1.771; p=0.077.

60 % would be satisfied with a charging time for 100 km of 30 minutes. 50 % would be satisfied with a charging time of one hour (cf. Figure 65).

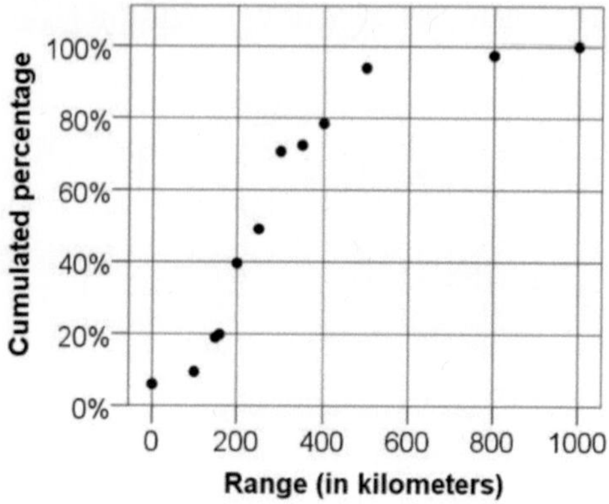

Figure 64: Minimum range of EV so that the EV users would consider purchasing an EV when they buy their next car (n=116).

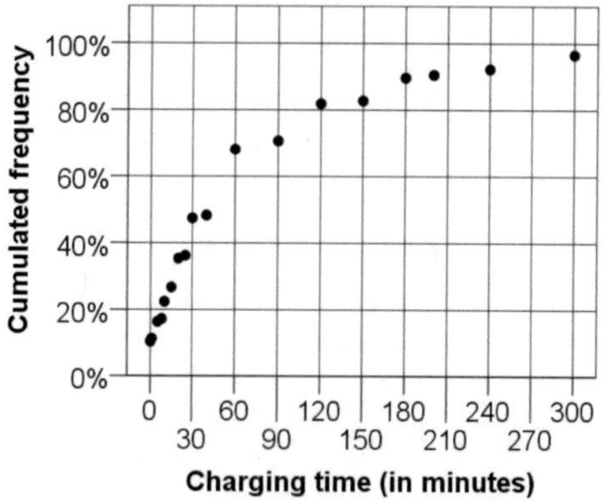

Figure 65: What is the maximum time an EV needs to be charged in for 100 km range so that you would actually consider an EV when you buy your next car? (n=116).

No significant differences between French and German respondents' answers concerning relative Willingness to Pay (WTP) for all kinds of PEV

(BEV, PHEV or REEV) were observed[55]. About 40 % of the EV users are willing to pay more for a BEV than for a gasoline driven ICEV. Somewhat less than 40 % state equal WTP between BEV and a gasoline driven ICEV (cf. Figure 66). WTP for a PHEV or REEV is somewhat higher. Somewhat less than 60 % of the EV users' state being willing to pay more for a PHEV or REEV compared to a gasoline driven ICEV. About 40 % of the EV users expect equal prices (cf. Figure 67).

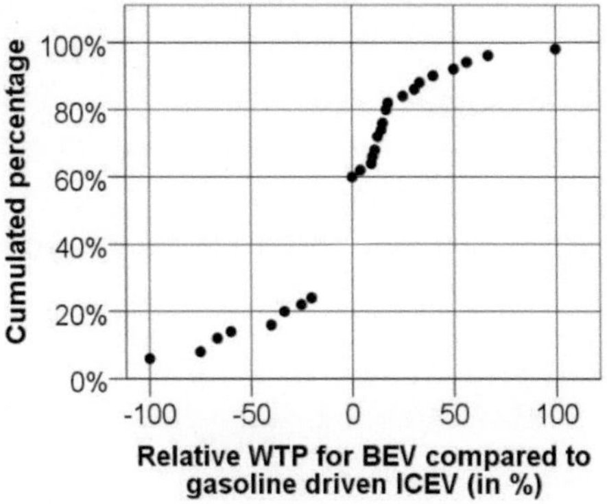

Figure 66: Relative willingness to pay (WTP) for BEV(n=50).

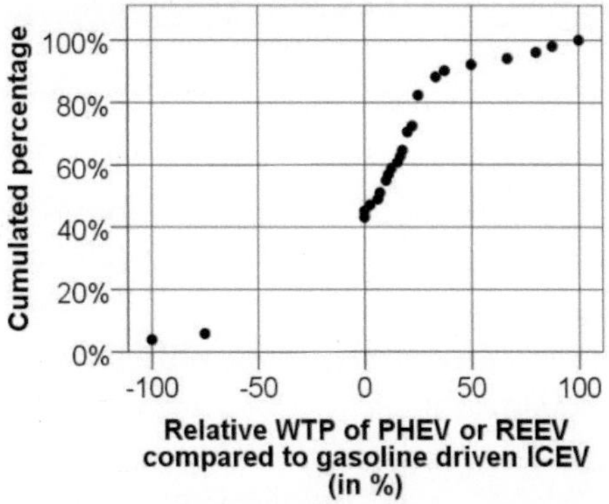

Figure 67: Relative willingness to pay (WTP) for PHEV and REEV (n=51).

[55] BEV vs. ICEV: N=50; Mann-Whitney U = 299; Z=-0.02; p=0.984.
PHEV vs. ICEV: N=51; Mann-Whitney U = 280.5; Z=-0.47; p=0.638.

8.10 Characterising EV adopters

In order to characterize the CROME users choosing a PEV during their next car purchase decision, the different powertrain alternatives are grouped in two clusters: A PEV cluster consisting of the two subgroups BEV and PHEV / REEV and another cluster including all other alternatives. According to Backhaus et al. (2008) a subgroup used in logistic regression analyses should at least contain 25 respondents. With regards to contents as well as limitations concerning the size of subsamples provided, this was the only possible option for pooling the subsamples, as the subgroup of respondents indicating EV choice only contained 25 users. In order to identify and characterise the EV users who stated that they will choose an EV during their next car purchase decision, a binary logistic regression model is estimated (n=86).

Table 13: Quality criterions of the Logit model.

Quality criterions for the model	Cox & Snell R^2: 0.404	Nagelkerke R^2: 0.577	Hosmer-Lemeshow test: p=0.565
Interpretation (cf. Backhaus et al. 2008)	Portion of variance explained by the predictors: (Very) good results		No significant difference between the forecasted values of the model and the observed values could have been observed. According to Backhaus et al. (2008) acceptable if p>0.7.

The model's parameters and corresponding interpretations indicate a good model fit particularly concerning the model's Nagelkerke and Cox & Snell R^2. According to the Hosmer-Lemeshow test results no significant difference between the forecasted values of the model and the observed values can be observed. Nevertheless, according to Backhaus et al. (2008) the p-value in this test should be above 0.7. Unfortunately this is not the case. An overview of the results is provided in Table 13. The following two equations describe the resulting probability of individuals' intention to choose a PEV during the next car purchase decision of the EV users depending on their car usage frequency, EV usage frequency, nationality, environmental awareness and willingness to seek for information on further development of EV in the future (cf. equations 1 and 2).

$$P(y = 1) = \frac{1}{1+e^{-z_k}} \quad (1)$$

with

$$z_k = -9.452^{**} - 2.741x_1^{**} + 2.508x_2^{*} + 2.002x_3^{**} + 1.219x_4^{**} + 0.552x_5^{*} + \varepsilon \quad (2)$$ [56]

Description of the variables:

y: Choice of a PEV when buying the next car (0: No / 1: Yes)

x_1: Daily car usage (0: No / 1: Yes)

x_2: Daily EV usage (0: No / 1: Yes)

x_3: Nationality (0: German / 1: French)

x_4: Information seeking on further development of EV (6 point Likert scale)

x_5: Environmental awareness (6 point Likert scale)

ε: Error variable

According to Table 14 all variables in the model discriminate significantly between the EV users choosing an EV when they purchase their next car and those who rather chose another powertrain alternative. Daily car usage negatively impacts the choice of a PEV in the next car purchase decision. On the other hand a daily EV usage has a positive impact at a similar magnitude. The French respondents are rather willing to choose a PEV as well as respondents who are environmentally aware[57] and who seek for

56 Significance level of Wald statistic: *: p<0.05; **: p<0.01; ***: p<0.001. Further information on the model's variables and coefficients is provided in Table 14.

57 Environmental awareness is measured by the item "If we continue as before, we are heading towards an environmental disaster". A 6 point Likert scale with an additional „Don't know" answering option is applied.

information on further developments of EV[58]. These results are in line with the findings of Frenzel et al. (2015) who interviewed more than 3,111 private as well as professional EV users. According to their results the motivation of German EV users to adopt EV can amongst others be explained by their interest in innovative technologies and their willingness to reduce their personal environmental impact.

Table 14: Quality criterions of the Logit model.

	β	S.E.	Wald	df	Sig.	Exp(β)	95 % C.I. for EXP(β)	
							Lower	Upper
x_1: Daily car usage	-2.741	0.821	11.131	1	0.001	0.065	0.013	0.323
x_2: Daily EV usage	2.508	0.967	6.725	1	0.010	12.278	1.845	81.709
x_3: Nationality	2.002	0.767	6.813	1	0.009	7.401	1.646	33.266
x_4: Information seeking on further development of EV	1.219	0.429	8.093	1	0.004	3.384	1.461	7.837
x_5: Environmental awareness	0.552	0.277	3.969	1	0.046	1.737	1.009	2.991
Constant	-9.452	2.816	11.266	1	0.001	0.000		

8.11 Conclusion and outlook

In this chapter, selected results of the third and final questionnaire among professional users of the CROME project were presented, notably the participants' socio-demographic backgrounds and their mobility behaviour including EV usage patterns. The comparison of usage patterns between French and German users showed that the German field test participants share their EV with more other users than the French. The French participants on the other hand use the EV in more diversified ways. The partici-

[58] Information seeking on further development of EV is measured by the item "I will keep myself informed about the future developments of EV". A 6 point Likert scale with an additional „Don't know" answering option is used.

pating fleets were examined including expectations of fleet managers concerning future fleet composition regarding powertrain technologies. According to the fleet managers' expectations the share of alternative powertrains will more than double until the year 2020. In the participating French organisations, the share of EV in the vehicle fleets is significantly higher already today – as well as in 2020 – than in the participating German organisations. Criteria influencing car purchase decisions of fleet managers purchasing cars for organisations and professional EV users envisioning to buy EV in the private context are compared. Several differences are observed, but criteria representing monetary aspects (purchase price) turned out to be the most important.

The next car to be purchased by the EV users will predominantly replace another car in the household. According to the respondents' answers, 27 % of them can envision purchasing an EV within the next 10 years. The observation of EV specific barriers showed that about 40 % of the EV users expect the prices of EV to be equal to those of ICEV, so they would consider purchasing an EV when they purchase their next car. On the other hand, 40 % of the survey participants providing answers on their WTP for BEV and ICEV would be willing to pay more for a BEV. Even 60 % would be willing to pay more for PHEV or REEV than for ICEV. National differences between French and German participants were examined concerning charging time. French users are suspected to wish slightly shorter charging times. 60 % of the respondents stated to be satisfied with a charging time of 30 minutes for 100 km. Concerning the range of the EV, 70 % of the users would be satisfied by a range of 300 km. Given this information about EV specific barriers, one can assume the potential that lies in further improvements, particularly concerning EV range and purchase prices. Potentials for EV adoption could certainly be increased by reducing these EV specific barriers. Binary logistic regression analysis was applied in order to characterise potential future private PEV adopters. The following two clusters of powertrain choice (1) PEV (BEV, PHEV, REEV) and (2) others (Hybrid, LPG/CNG, ICEV Gasoline, ICEV Diesel, Indifferent) were used. While daily car usage in general is assumed to constrain a positive choice of an electric powertrain, the model implies positive impacts of daily EV usage. Experiencing EV on a daily basis might lead to the conclusion that EV fulfil daily mobility needs. On the other hand, survey participants using conventional cars every day without making the experience of using an EV every day might not experience that an EV fulfils daily mobility needs. Further items positively impacting stated EV purchase intentions are environmental awareness as well as the willingness to seek for further information on the development of EV. On top of that, the model depicts the French users to be more likely to decide purchasing an EV in their next car purchase decision. As the sample sizes of participating fleet managers to assess future professional vehicle purchase decisions were limited, further data needs to be collected in order to obtain stable results.

8.12 Acknowledgements

This research was made possible by the CROME project [ref. no. 01ME12002], funded by the Federal Ministry for Economic Affairs and Energy and the Federal Ministry of Transport and digital Infrastructure in Germany and the French program "Investissements d'avenir programme véhicule de futur". Furthermore we would like to thank Magali Pierre (EDF R&D) and Anne-Sophie Fulda (EIFER) for their comprehensive help during the design and development phases of the three online questionnaires within the CROME project. We would like to thank Magali Pierre for the helpful comments for this chapter.

8.13 References

Backhaus, K., Erichson, B., Plinke, W., Weiber, R. (2008). Multivariate Analysemethoden. Heidelberg.

CROME (2014). Report on the Research Project CROME. http://crome-projekt.de/uploads/media/CROME_Report_December_2013_FINAL.PDF.

Dudenhöffer, K., Arora, R., Diverrez, A., Ensslen, A., Jochem, P., Tücking, J., 2014. Potentials for Electric Vehicles in France, Germany and India. In: Working Paper Serien in Production and Energy, Nr 5, ISSN: 2196-7296, http://www.iip.kit.edu/downloads/20141121_Dudenhoeffer_et_al_final % 281 %29.pdf.

Ensslen, A., Babrowski, S., Jochem, P., Fichtner, W. (2012). Existe-il des différences d'acceptation des véhicules électriques entre la France et l'Allemagne? – Premiers résultats de l'analyse scientifique du test de flotte Cross Border Mobility for Electric Vehicles (CROME). In: Proceedings of the 11ème séminaire francophone est-ouest de socio-économie des transports, Karlsruhe.

Ensslen, A., Jochem, P., Schäuble, J., Babrowski, S., Fichtner, W. (2013a). User acceptance of electric vehicles in the French-German transnational context. In: Selected proceedings of the 13th WCTR, Rio de Janeiro.

Ensslen, A., Jochem, P., Fichtner, W. (2013b). Experiences of EV Users in the French-German Context. In: Proceedings of EVS27 Conference, Barcelona, Spain.

Ensslen, A., Paetz, A.-G., Babrowski, S., Jochem, P., Fichtner, W. (2015) On the road to an electric mobility mass market – How can early adopters be characterised? In: Fornahl, D., Hülsmann, M. (Eds.), Markets and Policy Measures in the Evolution of Electric Mobility.

Eurostat (2012). National Accounts (including GDP). http://ec.europa.eu/eurostat/web/national-accounts/data/database. Accessed 17.03.2015.

Frenzel, I., Jarass, J., Trommer, S., Lenz, B. (2015): Erstnutzer von Elektrofahrzeugen in Deutschland. http://www.dlr.de/vf/Portaldata/12/Resources/dokumente/projekte/pakt2/Ergebnisbericht_E-Nutzer_2015.pdf.

Gagnol, P., Jochem, P., Pierre, M., Fichtner, W., 2013. „CROME: the French and German Field Demonstration of the Interoperable Mobility with EV". In: Proceedings of EVS27-Conference, Barcelona, Spain..

Globisch, J., Schneider, U., Dütschke, E., 2013. Acceptance of electric vehicles by commercial users in the electric mobility pilot regions in Germany, ECEEE Summer Study Proceedings.

Nationale Plattform Elektromobilität (2013). Fortschrittsbericht 2014 – Bilanz der Marktvorbereitung. https://www.bmbf.de/files/NPE_Fortschrittsbericht_2014_barrierefrei.pdf

Paetz, A.-G., Jochem, P., Fichtner, W. (2012). Driving electric vehicles in urban areas – comparing pre- and post-evaluations within the MeRegioMobil field test; mobil.TUM 2012, München.

Peters, A., de Haan, P. (2006). Der Autokäufer – seine Charakteristika und Präferenzen. Ergebnisbericht im Rahmen des Projekts „Entscheidungsfaktoren beim Kauf treibstoff-effizienter Neuwagen". Berichte zum Schweizer Autokaufverhalten Nr. 11 (in German). ETH Zurich, IED-NSSI, report EMDM2024, 39 pages (download from http://www.nssi.ethz.ch/res/emdm/).

Pfahl, S., Jochem, P. Fichtner, W., 2013. When will Electric Vehicles Capture the German Market? And Why?, In Proceedings of EVS27 Conference, Barcelona, Spain.

Pfriem, M., Gauterin, F. (2013). Less Range as a Possible Solution for the Market Success of Electric Vehicles in Commercial Fleets. In: Proceedings of EVS27-Conference, Barcelona, Spain.

Pfriem, M., Gauterin, F., Meyer, T. (2014). Selected Results from a Large-Scale Field Operational Test with Electric Vehicles in Germany and France. In: Proceedings of the 5th Hybrid and Electric Vehicle Conference. London.

Pierre M. (2014). Quand préoccupations environnementales et enjeux professionnels s'entrecroisent: le cas des entreprises participant aux essais de voitures électriques, in P. Lejoux et N. Ortar (Eds.), SHS Web of Conferences, Vol. 9 (2014), La transition énergétique : vrais enjeux, faux départs?

Pierre M. (2015). Script technique et prescriptions organisationnelles. Utiliser un véhicule hybride rechargeable en milieu professionnel, in Cihuelo J., Grandclément C. et Jobert A., Energie et transformations sociales. Enquêtes sur les interfaces énergétiques, éditions EDF R&D Lavoisier, Paris.

Plötz, P., Gnann, T., Kühn, A., Wietschel, M. (2013). Markthochlaufszenarien für Elektrofahrzeuge. Fraunhofer Institute for Systems and Innovation Research (ISI). http://www.isi.fraunhofer.de/isi-wAssets/docs/e/de/publikationen/Fraunhofer-ISI-Markthochlaufszenarien-Elektrofahrzeuge-Langfassung.pdf.

Schäuble, J., Ensslen, A., Jochem, P. Fichtner, W.: Electric vehicle market potentials- User acceptance, total costs of ownership, and changes in mobility behaviour. In: Fornahl, D., Hülsmann, M. (Eds.): Electric Mobility Evolution: Theoretical, Empirical and Political Aspects, Springer, Heidelberg, Germany (forthcoming).

Stella. K., Wollersheim, O., Fichtner, W., Jochem, P., Schücking, M., Nastold, M., Ensslen, A., Wietschel, M., Held, M., Gnann, T., Friedmann, M., Graf, R., Wohlfarth, K. (2015). Über 300.000 Kilometer unter Strom : Physikalisch-technische, ökonomische, ökologische und sozialwissenschaftliche Begleitforschung des Schaufensterprojektes RheinMobil: Grenzüberschreitende, perspektivisch wirtschaftliche elektrische Pendler- und Dienstwagenverkehre im deutsch-französischen Kontext. Karlsruhe. http://dx.doi.org/10.5445/IR/1000048691

Tanguy, F., Ayed, H., Baron, S., Berkel, F., Caspar, G., Dröschel, B., Glauer, G., Görges, D., Khadraoui, D., Tänzer, G. (2015): ELEC'TRA Endbericht. Metz, http://www.izes.de/cms/upload/publikationen/EM_12_091_D.pdf.

9 Quantitative analysis of public charging station use

Elise Nimal, Anne-Sophie Fulda

European Institute for Energy Research

Restricted EV's range is seen as a main drawback of electric mobility, leading one user to depend on an intermediary charge while a long distance travel is realized (Ministère de l'écologie du développement durable et de l'énergie, 2014). Therefore, convenient and reliable public charge may be considered as a catalyst for electric vehicles spread (Glachant, M., Thibault, M-L., Faucheux, L. 2011). Public decision makers are questioning about charging infrastructure: is a public charging station economically profitable? Where are the best locations? What are the targeted users? (Setec its, 2013).

During the CROME project, a lot of information about the electric vehicle charging behavior was generated. The interviews conducted in France have shown that users change their charging habits throughout the experiment (to a reasoned load) and that charging at home is very well accepted, more than at public infrastructure (cf. Chapter 4). This chapter adds information by focusing on feedback from the experiment on public charging station use in France. This is to better understand the practices of users and their appropriation of public infrastructure. The main questions discussed below are as follows: are stations used regularly? Does their location affect their use? Who are the most regular users? Are they private or professional users? How frequently do they charge the vehicle?

In order to answer these questions, we analyse the data gathered by the charging stations installed during the CROME project in France, and the files relating to the charging cards (badges) given to French participants in the project. The chapter is divided into two parts: analysis of the station use data and analysis of user behaviour, based on the records from the badges.

9.1 Station use

9.1.1 Charging stations' locations

The following paragraph describes the location of the stations which were analysed. There are nineteen stations located throughout Strasbourg region, five in the Colmar area and the remainder nine are localized in the east of France (Haut-Rhin, Bas-Rhin and Moselle). Figure 68 shows the charging stations located in the Strasbourg and Colmar city centers (whilst other stations - Forbach, Sarreguemines, etc. – are included in the data tables, for readability reasons they do not appear on the maps).

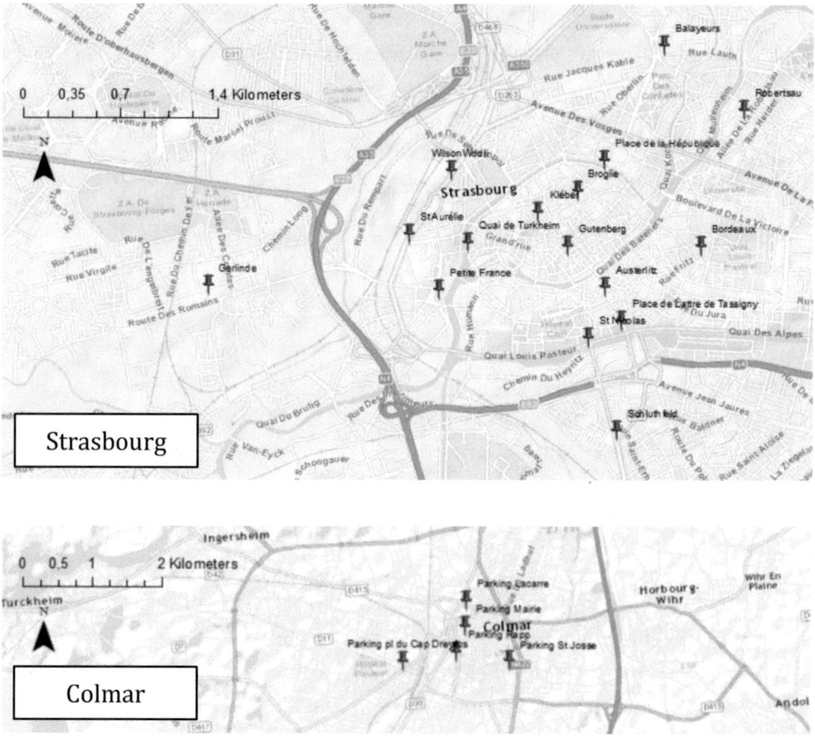

Figure 68: Location of charging stations, Colmar and Strasbourg city center (ArcGIS Google street base map).

There are 3 different types of stations:

- 5 fast charging stations close to motorway exits or to express ways in Cora supermarket car parks.
- 16 stations in car parks (semi-public, mode 2/3)
- 12 on the curbside (public, mode 2/3).

9.1.2 Database

First of all, we describe the data structure for the charging stations. The records are divided into three distinct tables, depending on the type of station they come from:

- Fast-charging stations (AC, DC) located in Cora supermarket car parks. The data was collected between April 1st, 2013 and 27 October 27th, 2014; this table consists of 2509 entries.
- Stations located in car parks and offering standard AC charging. The data was gathered between August 19th, 2013 and November 4th, 2014. The sample consists of 717 entries.
- On the curbside stations, offering accelerated AC charging. 6241 entries covering the period from April 1st, 2013 to October 27th, 2014. The RFID cards were identified for every charge at these stations.

Not all of these entries could be used and some are related to events which did not result in a charge. A filter was therefore introduced in order to retain only data which led to a charge. Ineffective charges are records considered as not resulting in batteries being charged; in this analysis, these are charges with a duration of less than 3 minutes and/or with zero energy consumption. Ineffective charges represent approximately 30% of the entries and they were removed from the subsequent analyses (Figure 69).

One possible explanation for the high number of ineffective charges is the lack of knowledge about how to use the stations; difficulties encountered when new vehicles (sometimes prototypes) first came into using public infrastructures may be another reason. Finally, it also appears that some stations did not communicate properly.

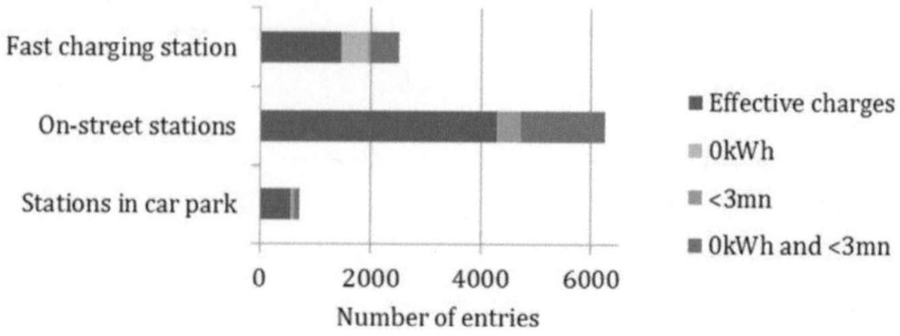

Figure 69: Spread of entries for all types of stations.

Manufacturer data, specific to station breakdown, was analysed in order to identify malfunctioning stations. Figure 70 shows the number of entry errors ("uncommunicated") per month and per station. "Uncommunicated" means: the charging station is working but it cannot communicate with the server. The number of errors of this type is very high representing more than 15483 entries (89% of the errors).

The number of "uncommunicated" errors is more or less evenly spread over all months in 2013 and 2014. All the curbside stations were affected in the same way by this communication error, though the Sarreguemines (Pax-Europole) station was more seriously affected than the others.

9.1.3 Charging frequency of stations

In the figures below, the number of effective charges is analysed per station in order to study its frequency of use. According to the number of entries reported by the car park stations, the latter are rarely used, with an average of 0.1 use per day since the stations were put into service. This frequency equals an average of one use per ten days.

Figure 71 shows the number of effective charges for each curbside station for the period April 2013 to October 2014. The average number of charges per day since the stations were put into service is 0.45 (meaning less than one use per two days), but this dissimulates significant differences ranging from 1.86 to 0.3 (excluding the defective Sarreguemines Pax-Europole station). The station located at Place de Lattre de Tassigny was identified as being the most frequently used: almost twice a day.

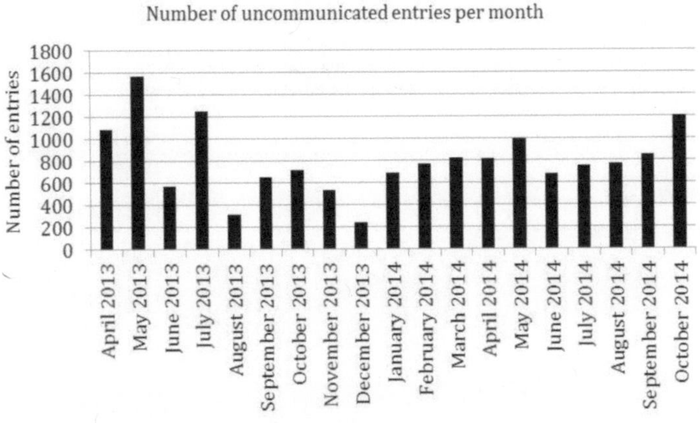

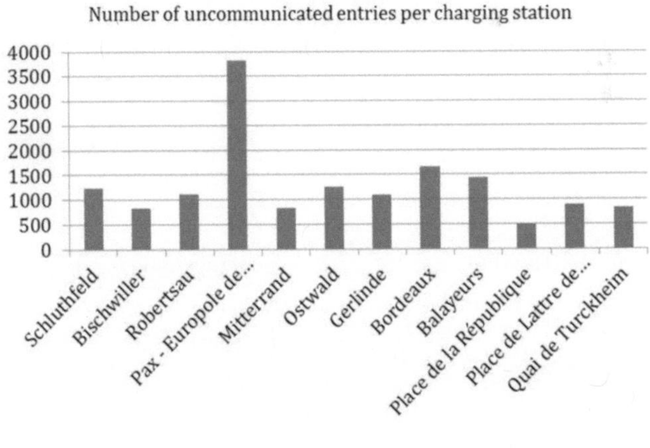

Figure 70: Number of occurrences with "uncommunicated" status for curbside stations per month (above) and per charging station (below).

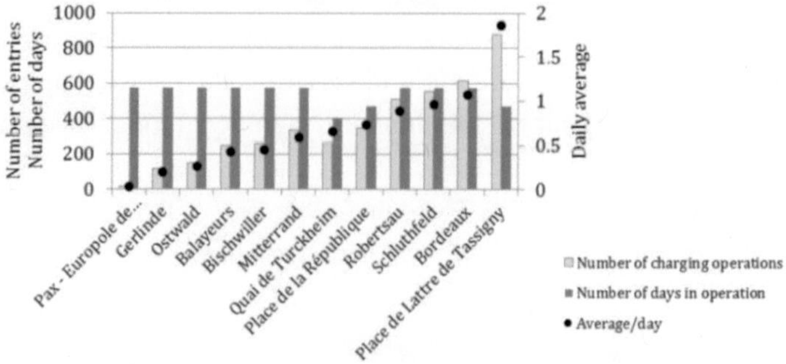

Figure 71: Number of days of operation since installation, number of charges and average number of charges per day at the curbside charging stations.

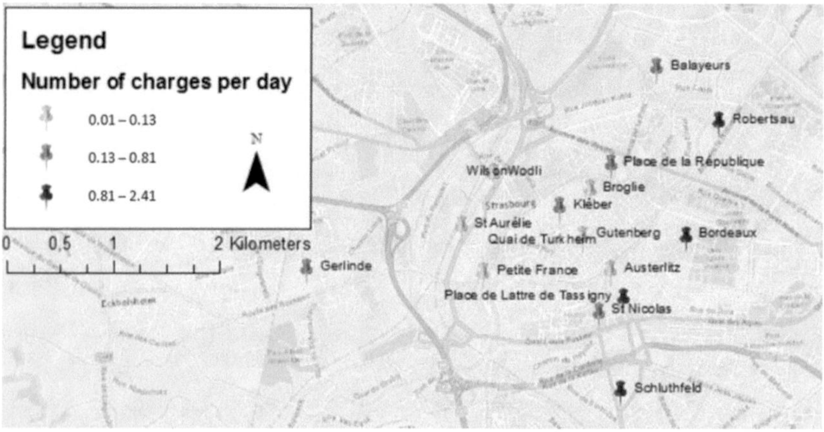

Figure 72: Number of charges per day for the Strasbourg curbside and car park stations (ArcGIS Google street base map).

Figure 72 describes the number of charges per day for each station in Strasbourg city center (curbside and car park). The stations located on the outskirts of Strasbourg city center are frequently used, whereas those located in city center car parks are less used.

The districts with the most frequently used stations are mixed and dense in nature. This might be explained by the users' activities as they perform a range such as work, shopping or leisure while they park for charging. They might also be local residents using the parking space (and the opportunity to charge) to avoid parking difficulties at home - this being a considerable problem in the investigated districts.

The stations located in car parks are rarely used. This might be due to their less efficient technology, but also to the difficulty of accessing them, as not all users necessarily have a car park pass.

Analysis of the data allows us to focus on the use of fast-charging stations. On average they were used for 1.3 charges per day since installation. Again, we can see a significant difference between the Molsheim station, which is the most frequently used with 2.15 charges/day, and the Strasbourg Sud station, with only 0.3 charges per day (Figure 73). Unfortunately, the sample is too small to allow further interpretation.

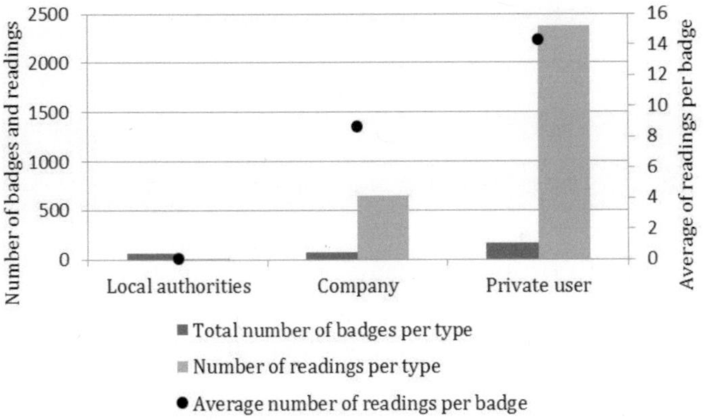

Figure 73: Number of days used since installation, number of charges and average number of charges per day for fast charging stations.

More specifically, for the period from October 17th, 2012 to February 15th, 2014, the data per station type show that activity is essentially concentrated at curbside stations and fast charging stations.

9.1.4 Charging time spreads

In order to better understand how the stations are used, we look at the weekly and daily spreads. Figure 74 shows that weekly use is almost the same for all types of station. The time-spread curve shows a higher level of activity between Tuesday and Friday. On Saturdays, fast charging stations are more likely to be used. On Sundays there is a clear drop in the number of uses for all stations.

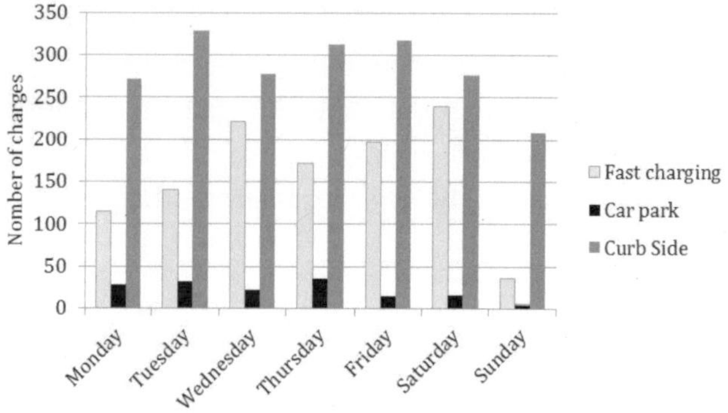

Figure 74: Weekly usage pattern of station type.

Figure 74 shows a weekly distribution of the number of entries between April 2013 and February 2014, for all stations in each category. The values are then aggregated by day of the week, regardless of the week or year in which the record was obtained.

The spread of charges per day is more uneven when shown per station type. The three graphs show the spread of charging start times throughout the day. On the one hand, the fast-charging stations are used throughout the day, with a peak at around 16h00.

The charging stations in car parks, on the other hand, are mostly used in the early morning; probably because this is the time that holders arrive to use the car parks while they are at work. Besides, the curbside stations are used throughout the day but especially on midday.

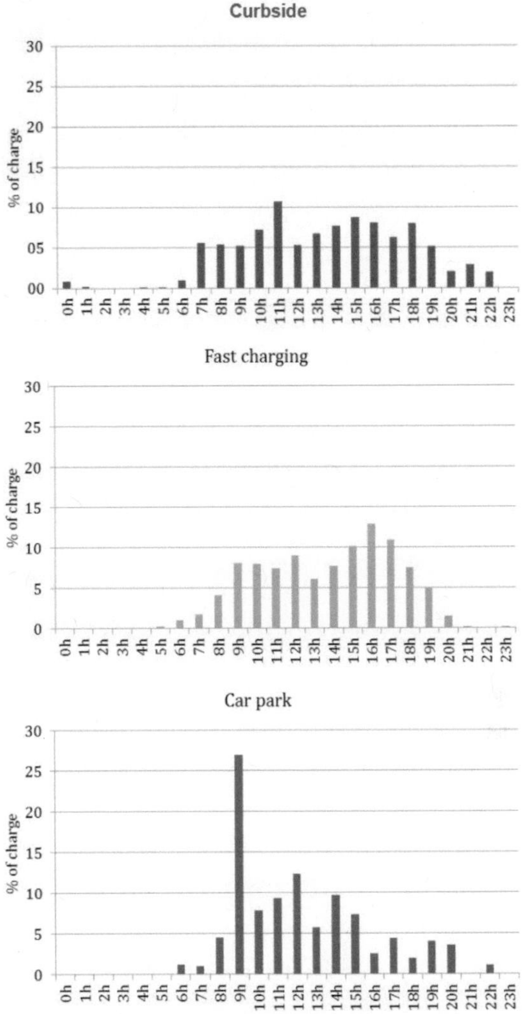

Figure 75: Daily charge spread (in %) depending on starting time.

9.1.5 User's preferences for charging stations

This paragraph analyses the data of badges used at the stations between April 2013 and October 2014. The analysis covers exclusively the curbside stations, as they are the only ones recording such information. The number of users varies at any given station.

At the Place de Lattre de Tassigny station for example, one badge holder charged his vehicle more than 300 times. It is nevertheless possible that this badge might be shared (a company badge, for example) and therefore corresponds to operations of several different persons.

Another station profile shows the Place de la République station: among the 45 badges identified, the maximum number of charges is 50 for an individual badge.

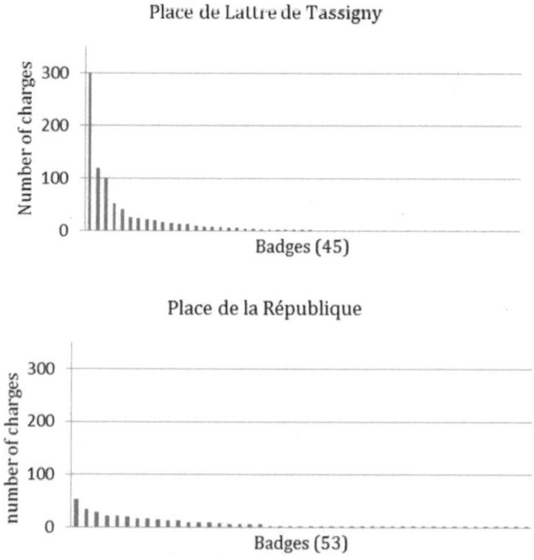

Figure 76: Number of charge per badge: example using two stations (Place de Lattre de Tassigny and Place de la République).

The regularity of the stations' usage is shown in Figure 77. This is the number of badges used per day at the station. Profiles vary significantly:

- The most frequently used station – Place de Lattre de Tassigny – is quite regularly visited by several users during the same day, with a few rare peaks and no real interruption.
- Stations which are less frequently used– Place de la République, Schluthfeld, Robertsau, Quai de Turckheim and Bordeaux (the example given below is based on data recorded at the Balayeurs station) – show both periods of non-use and periods of peaks.
- The remaining stations, rarely used (here is Ostwald station taken as an example), show an uneven profile, with numerous troughs and even some periods with no activity at all. The highest peak, with 6 badges read in one day, might be due to tests carried out by a maintenance agent.

To conclude, our analysis of the station data allowed us to identify differences in use between the various stations:

- Depending on type: available throughout the day, the curbside stations were the most frequently used.
- Depending on charge type: users fully appropriated the fast-charge stations.

Furthermore, certain stations are mostly used with a limited number of badges – albeit without being ignored by the other users – whilst other stations have multiple users but no predominant badge.

Finally, the daily rhythm varies with station accessibility (car park or curbside) and in accordance with charging type (fast, normal).

With a view to providing a more detailed understanding of how users accept public charging, the following section offers a "client analysis".

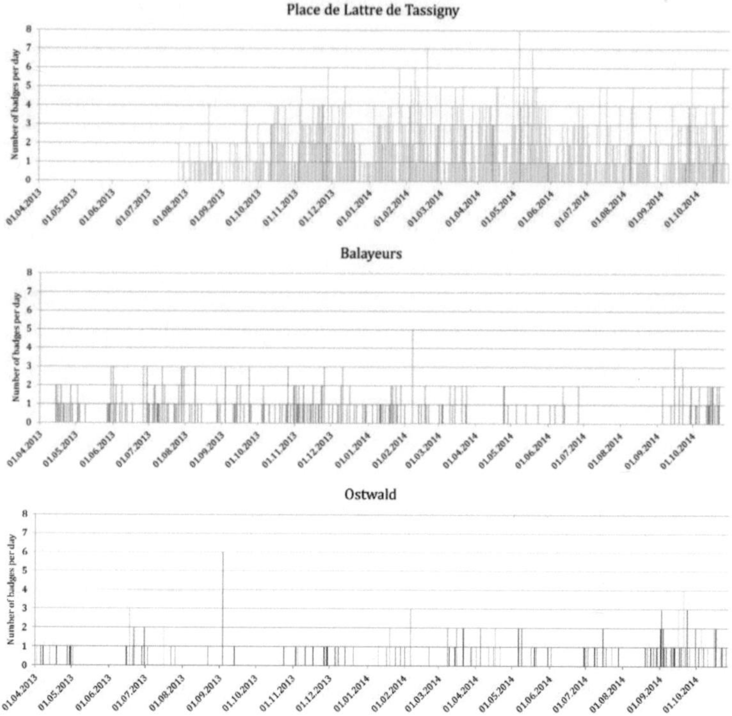

Figure 77: Number of usages per day with badge for three curbside stations: "place de Lattre de Tassigny", "Balayeurs" and "Ostwald".

9.2 User pattern

The database of badges managed by a French operator is described below.

In the badge file, received for feedback in the CROME project, users were recorded between 3 January 2013 and 4 September 2014. The analysed data base covers entries from:

- 264 badges,
- 237 different clients listed,
- and 299 entries.

The numbers of badges and clients differ due to multiple ownerships by people or companies.

The data base includes the badge identification number, the user type (local authority, private user or company), the badge owner's home location and, in some cases, the vehicle used by badge holders.

About a quarter of the badges are used by regional authorities, a quarter by companies and half by private individuals.

Only 100 of all badge holders provided details of the type of vehicle used. The Renault Zoé is the most recorded car, with 94 ownerships. Also listed in fairly significant numbers are Smart, Nissan Leaf, Kangoo ZE and Peugeot Ion. The fourth online questionnaire reveals that these vehicles are also the most popular with French respondents.

There are 132 users' name which do not relate to any particular type of car, either because the users did not wish to respond to the questionnaire or because a company owns several models and cannot predict which car will be used with a given badge.

9.2.1 Users' locations

Table 15: Number of badges per geographical zone and per type of badge holder.

Zone	Local authority	Company	Private	Total of entries
East (incl. Alsace and Lorraine)	63	68	163	294
Other locations		3	2	5
Zone total	63	71	165	299

294 of the 299 badges are referenced in the east of France. Table 15 shows that 63 badges are owned by local authority, 71 by companies and 165 by private users.

Figure 78 shows that the addresses linked to badges are spread throughout the east of France. Some addresses appear in the Moselle region, but the badges were mainly distributed in the Rhine Valley. Strasbourg and Colmar present a greater concentration of badges owners (due to higher population density). However, a large number of badges were also distributed in the more rural areas of western Alsace.

In Figure 79, the badges are classified in terms of user type. The map focuses on the urban area of Strasbourg, as this is the area with the greatest concentration of badges. The spread in terms of user type reveals no particular pattern.

The badge number is recorded in the databanks of curbside stations but not in the database from fast charging and car park stations. Consequently, the following paragraphs show an analysis of badges used exclusively at curbside stations.

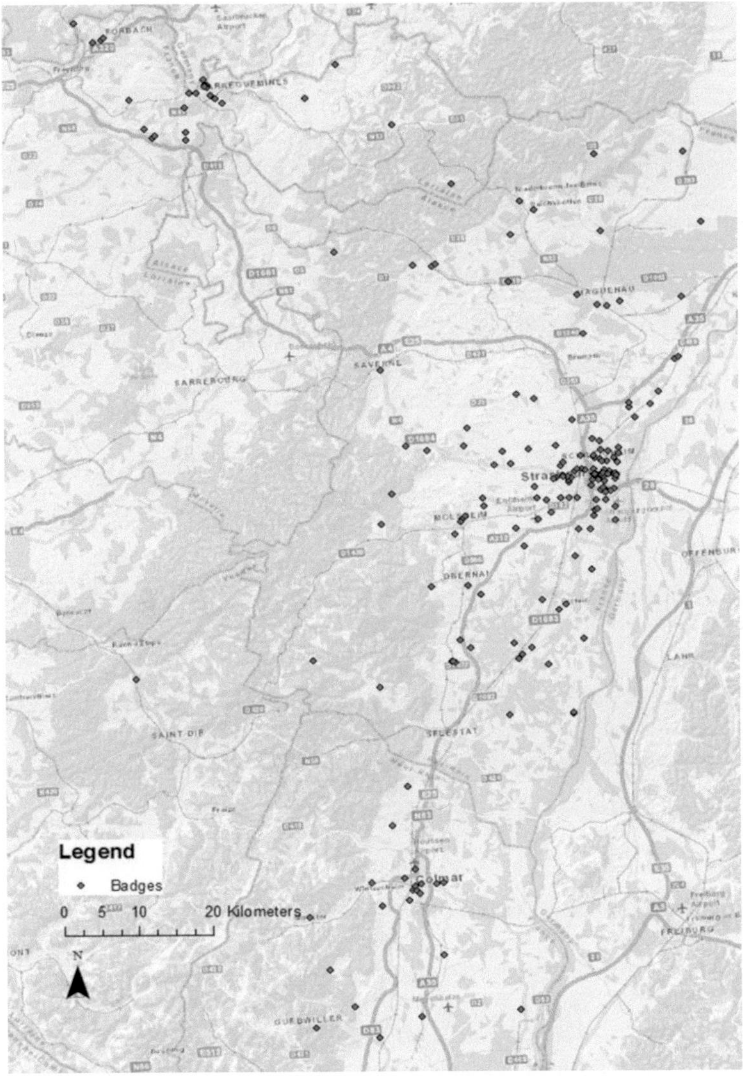

Figure 78: Badges' location in the east of France (ArcGIS Google street base map).

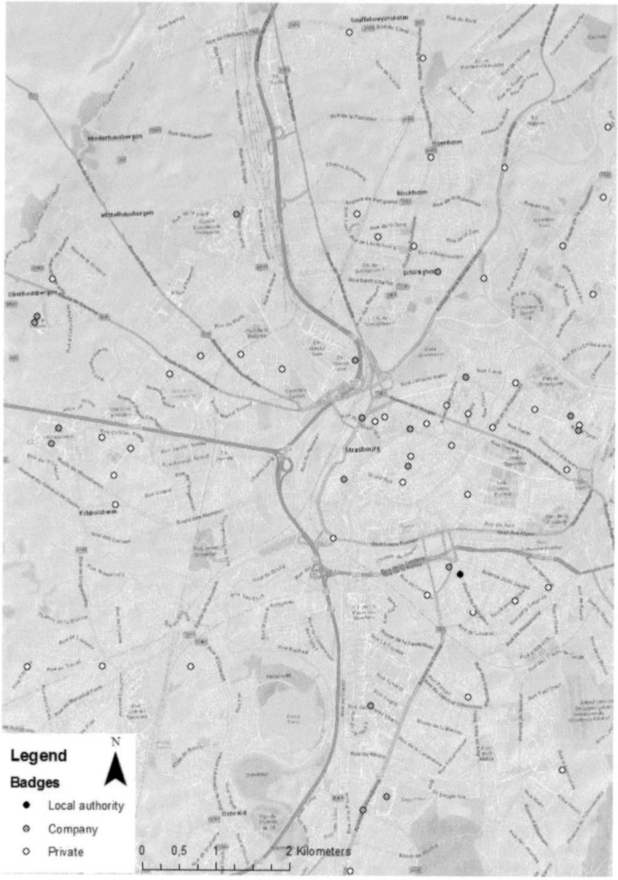

Figure 79: Badge location per user type, Strasbourg zone (ArcGIS Google street base map).

9.2.2 Number of stations used per badge

The following paragraph shows the number of stations used per badge. On average, a badge is used at 2.45 public stations but only 78 of 264 badges are used at more than one station.

Some badge holders use several charging points, which suggests a more flexible way of using the system. Some badges were used at many different stations (Figure 80); the vast majority of these belong to private users.

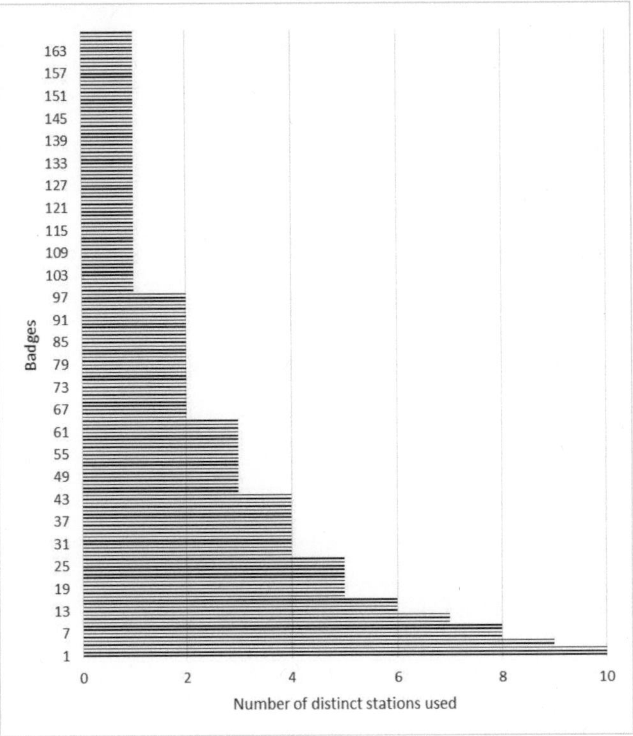

Figure 80: Badges classified in terms of stations used.

9.2.3 Frequency of badge usage

Details of badge use frequency are set out below, with the average number of charges, the differentiation of badge use in terms of user type and user typology.

On average, each badge was used 6.8 times at curbside stations during the period from 3 January 2013 to 4 September 2014. However, this average conceals a significant disparity in uses, particularly in relation to the type of badge holder (private user or company). Figure 81 shows the number of records listed in the tables for the curbside stations, per badge and user type.

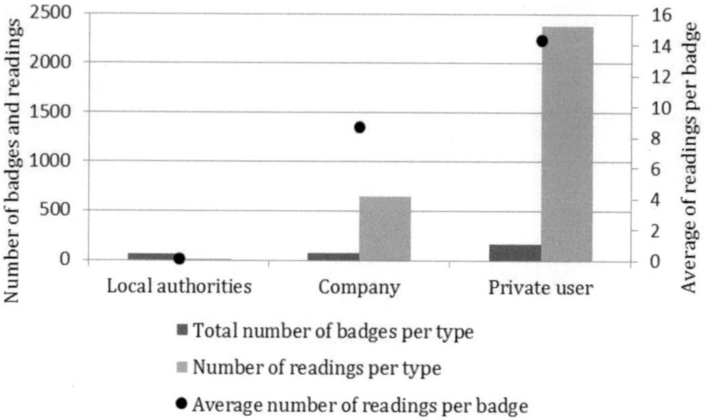

Figure 81: Relationship between number of badges and charges for local authorities, company and private user at curbside stations.

The majority of badges used at the stations relate to charges by private individuals; the latter are the most regular users, with an average of 15 charges per badge. Professional users charge less often – approximately ten registrations per badge – whilst public authorities hardly ever used their badges.

Figure 82 shows the number of charges at curbside stations, per badge. Each column relates to a badge. The number of uses describes an exponentially decreasing curve.

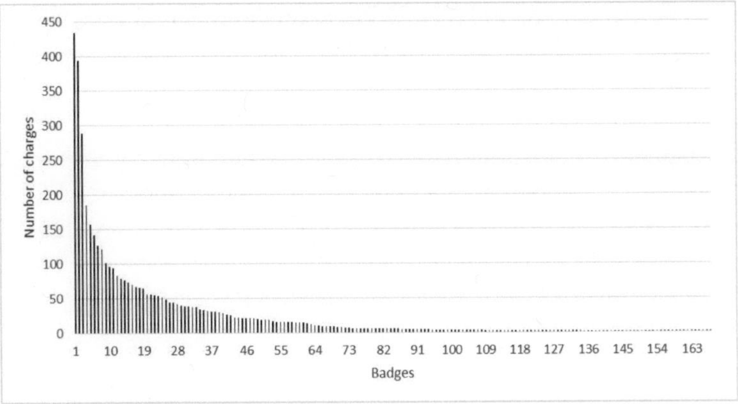

Figure 82: Number of uses per badge at curbside stations.

This distribution of the number of charges highlights a large number of badges with few uses, and a smaller number of badges from holders performing numerous charges. Eight badges were used more than one hundred times and only two of these belonged to companies. The lower number of uses may be due to the fact that they do not need to charge at public stations.

Finally, the badges were classified in terms of number of uses, using the Ascending Hierarchical Classification (AHC) method. Figure 83 highlights three types of behaviour:

- Category 1 corresponds to very infrequent users, possibly persons who wish to try a public charging station out of curiosity.
- Category 2 corresponds to relatively frequent users. These may be people who use public stations for additional charging; this result is suggested by a sociological survey on private and professional users (cf. Chapter 4).
- Category 3 corresponds to very regular charges in public stations. According to the qualitative survey, it may refer to constrained users who are unable to charge at home (Pierre, M., Jemelin, C., Louvet, N., 2007); this increased use may also relate to maintenance personnel.

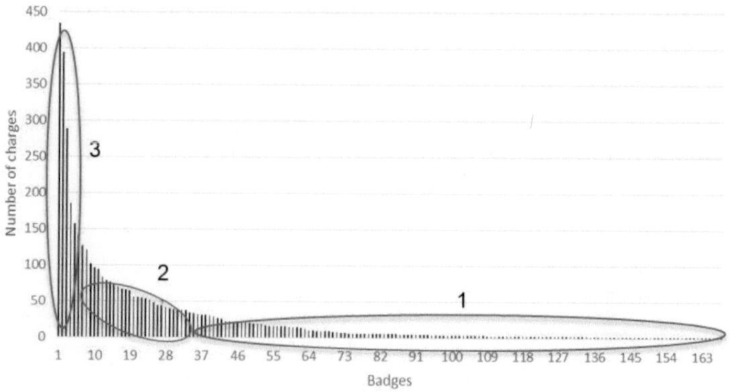

Figure 83: Number of charges per badge with user categories highlighted.

9.2.4 Location of badges and used stations

A cross-analysis of badge location and station use shows that behaviours vary considerably. Two types of user would appear to be particularly representative of private individuals:

- Firstly, certain users seem to rely on public charging stations and charge with very significant frequency at a single station located close to their homes;
- Secondly some users plug in at the city center, at various public stations. This analysis highlights the following typologies of use, interpreted as: "constrained charging" and "additional charging".

Finally, badges analyses showed that private individuals use the public station infrastructure more than professionals. The system is used by a relative small number of regular users who charge at different stations. Two user profiles are highlighted: users who consider stations as additional charging points, and constrained users who rely on the public charging infrastructure.

9.3 Conclusion

This chapter has demonstrated that public charging infrastructure is required for electric vehicles deployment; especially for users that depend on it or consider it as additional charging. Analysis of the charge and client data bases allowed us to make a differentiated appraisal of public charging.

During the CROME project, electric vehicle users were able to experiment with the public charging infrastructure; they experienced particular difficulties when using the stations, a certain number of which were out of order (Pax-Europole for example). A significant number of badges remain unused after the user tests maybe because they do not need public charging. However, some users readily appropriated public stations for additional charging. Certain stations - particularly curbside stations in the denser districts of Strasbourg – were frequently used.

Finally, there is the case of "constrained charging" for a small number of users who appear to rely on public charging stations because they are unable to charge elsewhere.

Users also seem to have appropriated the service available from fast-charge stations.

It would be useful if this analysis of feedback from the experiment will be an ongoing process, so as to gain a better understanding of user behaviour, for example by refining user profiles over a larger sample (in years to come) and through a more effective examination of fast-charge stations when it becomes possible to identify badges during the charging process.

9.4 References

Glachant, M., Thibault, M_L., Faucheux, L. (2011): Le déploiement des infrastructures de charge de véhicules électriques et hybrides rechargeables: une approche économique, Mines ParisTech, Armines, p.80.

Ministère de l'écologie du développement durable et de l'énergie (2014), Guide technique pour la conception et l'aménagement des infrastructures de recharge pour véhicules électriques et hybrides rechargeables. Mise à jour du volet technique du Livre vert sur les infrastructures de recharge ouvertes au public pour les véhicules décarbonés, p.36.

Pierre, M., C. Jemelin, N. Louvet (2011). Quel public et quels usages pour la voiture électrique? Enseignements issus des retours d'expérience conduits à EDF R&D, EDF, EPFL (Ecole Polytechnique Fédérale de Lausanne), p.50.

Setec its (2013). Déploiement des infrastructures de charge pour les véhicules électriques et hybrides rechargeable, p.35.

10 Integrating BEV into daily travel behaviour

Tim Hilgert, Martin Kagerbauer, Christine Weiss, Peter Vortisch

Institute for Transport Studies, Karlsruhe Institute of Technology

In addition to the investigation of BEV in business contexts, particular interests focus on how BEV can be used in people's everyday life. We investigated the possible usage of BEV for the CROME field study participants. Therefore we used national household travel survey data (MOP) and estimated the share of car trips that can be substituted by BEV. Our results show that the participants of the CROME field study have a car usage frequency and intensity above average. These participants can replace most of their trips by a BEV. However, substituting their existing conventional vehicles completely by BEV is in most cases not feasible since they cannot replace all their trips.

10.1 Introduction

As part of the CROME research project, data of BEV' usage patterns and surveys about the participants' behaviour were carried out. The participants of the project mainly used BEV of companies' car fleets for business trips. Based on this field study, the research of the Institute for Transport Studies at the KIT consists of two parts:

We analyse business car trips to show differences of the usage between trips made with BEV and conventional vehicles ICEV. Therefore, we use the data of different travel surveys. We investigate that the trips made with BEV during the CROME research project are shorter than comparable trips made with ICEV.

Another aspect deals with the usage of BEV in people's everyday life. For that reason, we model and analyse the daily travel behaviour of the CROME research projects' participants. Finally, we estimate the potential of integrating BEV in their daily travel behaviour. The results show that using BEV in a company's fleet is a realistic scenario. However, the potential of substituting a personal ICEV completely by a BEV is very low concerning this target group. In general, they have a car usage above average. That causes range problems when substituting all car trips by BEV.

10.2 Data and models used

For our analyses, we use the data of three surveys which are outlined in the following: the CROME field study, the German survey MOP and the French survey ENTD. In addition, we use the CUMILE model which bases on MOP data.

The data of the CROME field study involve trip data and data of participants carried out by online surveys during the project (cf. Chapter 1). The trip data consists of two sources. First, individual trips (begin, end, length, distance, etc.) gathered by data loggers are available. These data loggers were directly connected to the vehicles and collected the data automatically for every trip. Second, trip information gathered by smartphones which were attached to the vehicles are available (cf. Chapter 7). Altogether, we use 10,240 trips of data loggers, 6,418 trips collected by smartphones and survey data of 286 participants for our analysis.

Since 1994 the German Mobility Panel annually collects data on the travel behaviour of the German population. It is both – a multi-day and a multi-period travel survey. Each year approximately 1,000 households with 2,000 persons (10 years and older) participate to the MOP survey by filling in a travel diary in the course of one week. They provide detailed information on all trips including times, modes, purposes and distances. Participants also provide information on sociodemographic characteristics. The survey is carried out on behalf of and funded by the German Federal Ministry of Transport and Digital Infrastructure. The Institute for Transport

Studies at the KIT is responsible for the design, scientific supervision and analyses of the survey. (Streit et. al. 2014)

Our analyses are based on the MOP data collected from 2003 to 2012 with a sample of 9,016 participants. To estimate the group of business car trip makers, we use the data of 1,964 persons of the full sample who performed 8,504 business car trips.

The French survey ENTD is a one-day survey that gathered the daily travel behaviour of people in the years 2007/2008 with a sample of 2,026 business car trips (Shanti – EU Travel Survey Wiki 2013).

Additionally to the surveys, we use the CUMILE model to extend our researches. This model has been developed by the Institute for Transport Studies and depicts car mileages at every day of the year of the German car fleet (Weiss et. al. 2014). CUMILE is also based on the data of the MOP and several other surveys.

10.3 Methodology

In the following section, methods are presented for the analysis of business trips, the modelling of the daily travel behaviour and the integration of BEV.

10.3.1 Analysis of business trips

To analyse business trips, we use the data of all three mentioned surveys: the CROME field study, the MOP and the ENTD.

For the analysis of trips with BEV, the trip-data collected during the CROME field study is compared to other business trips with ICEV. The trip data for ICEV is gathered by the two other national surveys (MOP and ENTD). To show differences in the car usage, we compare the average trip length of business car trips. Therefore, we investigate confidential intervals of average trip lengths.

10.3.2 Modeling daily travel behaviour

At the first step of the travel behaviour's investigation, we compare the daily travel behaviour of business car trip makers between the CROME and the MOP participants. Since the CROME field study mainly contains business trips with BEV, the daily travel behaviour of the field study participant's remains unknown. Therefore, we developed a model that synthetically generates this assumed behaviour. This model is based on the data of the CROME field study and the MOP.

For this model, we implement a variation of the statistical matching method. The method of the statistical matching is often used in different contexts, e.g. market research. It helps to link different layers of information. In the field of the market research, information are often available for a subset of people and other information for another subset only. The statistical matching method helps to synthetically model all information for the whole set of people. (Noll 2009)

The surveys of the CROME project contain data of the participants' sociodemographic and the commuting habits. The overall travel behaviour of the CROME field study's participants is reproduced by using the CROME and MOP data. The matching is based on the assumption that the travel behaviour can be predicted by using statistic similarities of persons' sociodemographic and commuting information only.

With the method of the statistical matching, we compare different variables (age group, gender ...) of the CROME and the MOP participants. Statistical twins are identified in the set of the MOP participants according to selected variables. This allows us to estimate the unknown travel behaviour of the CROME participants. In order to find the best set of variables, we developed an algorithm that searched for the best matching. It iteratively performed the matching for every possible combination of variables.

Figure 84 shows an example of the statistical matching method.

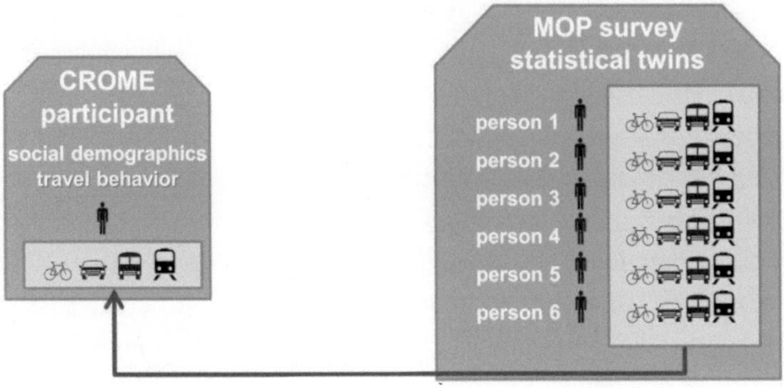

Figure 84: Example of the statistical matching method to reproduce the overall travel behaviour for a CROME participant.

In this example, the CROME participant has six statistical twins that were found in the set of the MOP. These six statistical twins represent the travel behaviour of the CROME participant. To determine characteristic travel values, e.g. the modal split, the values are weighted (1/6 in this case).

10.3.3 Integrating BEV

Our developed algorithm allows even to integrate BEV into the travel behaviour. It uses the modelled data for the CROME participants from the previous step as well as the travel behaviour data of business car trip makers of the MOP. The algorithm proves whether a trip could be done with an BEV or not. In considering all trips it is possible to define the potential of BEV for the CROME participants as well as for the business car trip makers of the MOP.

To make a trip with a BEV, the trip has to fulfil different criteria. The first criterion considers the mode of transport. We only substitute trips with cars and several trips with bikes or by foot. Other studies like Krems et. al. (2011) analysed a potential for substituting trips that were previously done with bikes or by foot and are now done with BEV. This effect was mainly caused by the environmental image of BEV. The second criterion considers the distance of a trip. It should not exceed half of the range of the BEV (including a comfort distance of the participant since we assume that the actual infrastructure does not ensure charging facilities at the final destination). As a third criterion, the trip has to be a part of a tour of trips that can be done with a BEV. Additionally to the idea of the distance criterion and the charging facilities, this criterion combines sets of trips to tours that begin at home and end at home. A tour can be done with a BEV if the range of the vehicle do not fall below 10% of the full range by performing this tour. Figure 85 demonstrates the principle of combining trips to tours for a period of one day.

Using this method, we estimate all tours and decided whether they are feasible for BEV or not. In this example, the third tour cannot be done with a BEV since it exceeds the actual possible range of the vehicle.

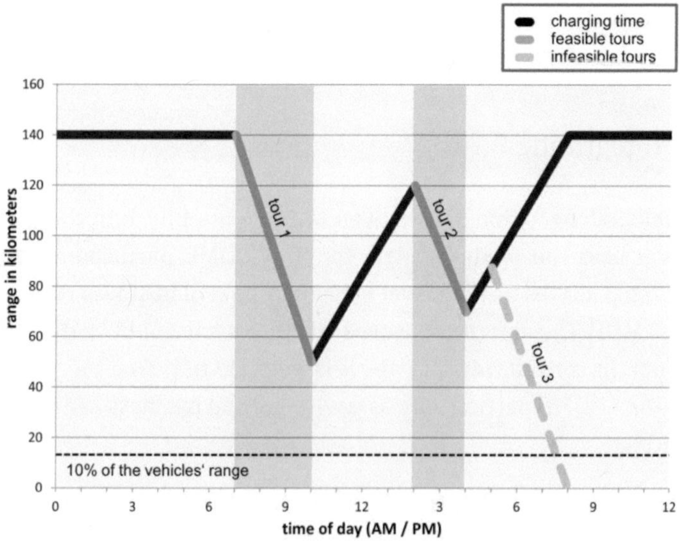

Figure 85: BEV range during a day to estimate potential tours for BEV.

Additionally to the consideration of one week based on the MOP data, we investigate the distance travelled at all days of the year using the CUMILE model. This model is also based on the data of the MOP but depicts car mileages at every day of the full year of the German car fleet. Because the model is based on the MOP we use the determined statistical twins of the CROME participants for the model. Subsequently we estimate the days of the year that exceed the range of the BEV for the CROME participants. In addition to an average week, we also consider seldom events in the travel behaviour such as holiday trips by using the CUMILE model. Furthermore, this implies two perspectives: One consideration based on an average week and on trip basis and one based on the period of a full year on daily basis.

10.4 Results

Results are presented for the analysis of business trips, the modelled daily travel behaviour without BEV and after integrating BEV.

10.4.1 Analysis of business trips

The trips gathered within the CROME field study are comparably short (cf. Chapter 7). Most of them are in a distance scope of 10 km. This is one of the major results and the most significant difference to trips that were gathered by other surveys. Considering the other surveys, business trips have an average length of 20 to 30 km. One reason for this difference might be the limited range of the BEV. The vehicles used in the CROME vehicles fleet had an average range of about 150 km. Regarding the infrastructure of charging possibilities nowadays, charging at the place of destination is not ensured. This is why the driver always has to consider the trip back in his mode choice decision. Considering these aspects, the maximum distance of a trip is about the half of 150 km.

To compare the trips between the different surveys, we estimated the 95 % confidential interval of the average trip length. For the surveys MOP and ENTD, we made two analyses: One without a constraint in distance and one with the constraint that trips longer than 77 km are not considered due to the aspect of the range. The following Table 16 shows the examined results.

Table 16: 95 % confidential interval of the average trip length for the surveys used.

Survey	Unconstrained	Trips ≤ 77 km
CROME (data loggers)	[6,5 ; 6,8] km	-
CROME (smartphones)	[8,7 ; 9,1] km	-
MOP 2003-2012	[24,2 ; 26,5] km	[13,8 ; 14,5] km
ENTD 2007/2008	[21,7 ; 25,5] km	[15,7 ; 17,1] km

Regarding the constrained trips to a maximum of 77 km, trips during the CROME project are still shorter than trips of other surveys with ICEV. These differences may have several reasons:

The vehicles fleet in the CROME project consisted mainly of small cars like the Smart ED. Cars in this class are normally used for shorter trips. Additionally to that, using a car in a vehicles fleet offers the possibility of choosing the right vehicle for a specific trip. The results of the user surveys during the project indicate that 79% of the vehicles used in CROME do not replace an existing vehicle of the company's fleet. If the driver knows that the range is not sufficient for an upcoming trip, he or she chooses an ICEV of the fleet. Subsequently, BEV are mainly used for shorter trips.

Another reason for shorter trips arises by substituting trips former done by foot or bike. This effect has also been examined in other studies like the MINI E study (Krems et. al. 2011). Caused by the positive image of BEV concerning environmental sustainability, people substitute some of their trips. Trips done by foot or bike are normally shorter than trips by cars. Consequently the share of shorter trips increases.

The last reason to explain the fact of shorter trips at the CROME project deals with the structure of the sample at CROME. There are several cars in the sample that have a high impact on the average trip length. As an example, one car logged over 2,000 trips with an average length of about 3 km. At other surveys like the MOP or ENTD, cases like this do not have such a high impact due to weighting factors that ensure the representativeness of the sample.

10.4.2 Analysis of daily travel behaviour

The results of the travel behaviour for the CROME participants, represented through the statistical twins, as well as for the business car trip makers of the MOP indicate a high use of cars in their daily travel behaviour. The share of car trips is above average. Table 17 shows the results for the share of trips done with cars and the distance travelled per week for the different groups. As a reference, the table also includes the share and the distance for all participants of the MOP (9,016 participants).

Table 17: Comparison of average car usage and average distance traveled per week of CROME and MOP participants.

Target Group	Average share of trips by car	Average distance travelled per week
CROME – participants	80 %	471 km
MOP – business car trip makers	71 %	520 km
MOP – all participants	41 %	282 km

The high share of car trips has several reasons. 71% of the CROME participants and 80% of the business car trip makers of the MOP use their car to commute to work. This fact causes that other trips are also often performed with cars due to convenience, e.g. shopping on their way back home. Second, CROME participants often use their car because of their place of living. Over 58% of the participants live in a rural area or in a small city. Usually, public transport is less developed in these areas.

There are also several reasons to explain the high distances travelled per week in addition to the reasons above. The CROME field study covers working people only. Regarding the average distance travelled of all persons, also the distances of for example children or seniors are taken into account. They often make trips with lower distances. Business car trip makers also have higher distances travelled per week because they often perform trips during their working time. While a lot of working people typically do not leave their working place during their working time, business car trip makers also make trips during their working time, for example visiting a client. This also causes a distance travelled per week above average.

10.4.3 Integrating BEV

We first show the results considering an average week of the year under the assumption that every person has a BEV available beside their ICEV. These results base on the developed algorithm presented in the methodology section. Using this algorithm we investigate every trip and decide whether it could be conducted by a BEV or not.

We estimate the share of BEV trips for the CROME participants (using the modelled data of the statistical twins) and figured out that no participant can completely replace his ICEV. On average, still 13% of the trips have to be done with ICEV. We also calculated the shares for the business car trip makers of the MOP and set two different BEV ranges (135km and 255km). The results are similar and mainly caused by the high distance travelled and the car usage above average shown in the previous section. Table 18 shows the results in terms of the modal split including BEV when they are available in addition to conventional vehicles.

Additionally, we analysed on how many days of a full year the daily car mileage exceeds 90 % of the BEV range. Therefore, we used the CUMILE model. Table 19 shows that the average number of days of the CROME participants exceeding the vehicles' range requirements is 35.

Table 18: Comparison of average shares of trips for different modes of transport for CROME and MOP participants.

Target Group	ICEV	BEV	Other modes
CROME – participants	13 %	72 %	15 %
MOP – business car trip makers (135 km range)	14 %	63 %	23 %
MOP – business car trip makers (255 km range)	10 %	67 %	23 %

Table 19: Number of days that exceed the BEV-range using the CUMILE model for the CROME and MOP participants.

Target Group	Average	25 %-Quartile	75 %-Quartile
CROME – participants	35 days	13 days	50 days
MOP – business car trip makers (135 km range)	41 days	6 days	55 days
MOP – business car trip makers (255 km range)	17 days	1 day	21 days

10.5 Conclusion

We analysed the travel behaviour of business car trip makers and focused on the potential to integrate BEV in the daily travel behaviour of vehicle users. Despite different methods combining statistical analysis and simulation, the results are consistent with each other.

Our results show that the car usage frequency and intensity of the group of business car trip makers is above average. This fact makes them interesting for the use of BEV in general. However, they cannot replace all conventional car trips by BEV since some tours exceed the BEV range. On the one hand, they can perform over 70% of the trips with BEV on average. On the other hand, there is nearly no potential to completely substitute existing ICEV.

Within the CROME project, BEV were used in company fleets. This is a realistic scenario and can be a valuable opportunity for using BEV. Being part of a fleet reduced the weakness of the limited range in a personal environment since other cars are available. BEV are a good possibility to cover the high share of tours with a sufficient distance range due to the possibility of choosing.

10.6 References

Krems, J.F., Bartholdt, L., Cocron, P., Dielmann, B., Franke, T., Henning, M.J., Ischebeck, M., Schleinitz, K., & Žilyte-Lennertz, M. (2011). MINI E powered by Vattenfall V2.0. Schlussbericht zum Forschungsvorhaben Verbundprojekt: MINI E powered by Vattenfall V2.0 (Abschlussbericht, Förderkennzeichen 16EM0070). Chemnitz: Technische Universität Chemnitz.

Noll, P. (2009). Statistisches Matching mit Fuzzy Logic Theorie und Anwendungen in Sozial- und Wirtschaftswissenschaften. Wiesbaden: Vieweg + Teubner.

Shanti – EU Travel Survey Wiki. (2013). France:Enquête Nationale Transports et Déplacements (ENTD); National Survey Transportations and Travel.

Streit, T., Chlond, B., Vortisch, P., Kagerbauer, M., Weiss, C., & Zumkeller, D. (2014). Deutsches Mobilitätspanel (MOP) – Wissenschaftliche Begleitung und Auswertungen Bericht 2012/2013: Alltagsmobilität und Fahrleistungen. Schlussbericht zum Forschungsprojekt FE-Nr. 70.0813/ 2007. Karlsruhe: Institut für Verkehrswesen des Karlsruher Institutes für Technologie.

Weiss, C., Chlond, B., Heilig, M., & Vortisch, P. (2014). Capturing the Usage of the German Car Fleet for a one Year Period to Evaluate the Suitability of Battery Electric Vehicles – a Model Based Approach. Transportation Research Procedia, Vol 1, p. 133-141